Use R!

Series Editors:
Robert Gentleman   Kurt Hornik   Giovanni Parmigiani

Use R!

Phil Spector

Data Manipulation with R

 Springer

Phil Spector
Statistical Computing Facility
Department of Statistics
University of Califonia, Berkeley
Berkeley, California 94720
spector@stat.berkeley.edu

Series Editors:

Robert Gentleman
Program in Computational Biology
Division of Public Health Sciences
Fred Hutchinson Cancer Research Center
1100 Fairview Avenue, N, M2-B876
Seattle, Washington 98109-1024
USA

Kurt Hornik
Department of Statistik and Mathematik
Wirtschaftsuniversität Wien Augasse 2-6
A-1090 Wien
Austria

Giovanni Parmigiani
The Sidney Kimmel Comprehensive Cancer
Center at Johns Hopkins University
550 North Broadway
Baltimore, MD 21205-2011
USA

ISBN 978-0-387-74730-9   e-ISBN 978-0-387-74731-6
DOI: 10.1663/978-0-387-74731-6

Library of Congress Control Number: 2008921862

Printed on acid-free paper.

9 8 7 6 5 4 3 2 1

springer.com

Preface

The R language provides a rich environment for working with data, especially data to be used for statistical modeling or graphics. Coupled with the large variety of easily available packages, it allows access to both well-established and experimental statistical techniques. However techniques that might make sense in other languages are often very inefficient in R, but, due to R's flexibility, it is often possible to implement these techniques in R. Generally, the problem with such techniques is that they do not scale properly; that is, as the problem size grows, the methods slow down at a rate that might be unexpected. The goal of this book is to present a wide variety of data manipulation techniques implemented in R to take advantage of the way that R works, rather than directly resembling methods used in other languages. Since this requires a basic notion of how R stores data, the first chapter of the book is devoted to the fundamentals of data in R. The material in this chapter is a prerequisite for understanding the ideas introduced in later chapters.

Since one of the first tasks in any project involving data and R is getting the data into R in a way that it will be usable, Chapter 2 covers reading data from a variety of sources (text files, spreadsheets, files from other programs, etc.), as well as saving R objects both in native form and in formats that other programs will be able to work with. Chapter 3 addresses the issue of relational databases, since large datasets are often stored in such databases. Some guidance in setting up and using databases to work with large datasets is also included in this chapter.

Chapter 4 covers the topic of dates and times in R. While some work can be done using a simple character representation of this type of data, a wider range of operations are available when dates and times are converted to an internal form that allows for comparisons and other manipulations. There are a variety of mechanisms for storing dates and times in R, and this chapter is presented to encourage users of such data to convert them to the appropriate type as early as possible.

While factors are undeniably valuable in data modeling and graphics, they often "get in the way" when performing more basic operations on data. Chapter 5 addresses how to convert objects to and from factors, along with guidelines on how to avoid factor conversions when necessary.

Chapter 6 explores the many ways that subscripting in R can be used to access and modify data. Subscripts (especially logical subscripts), are one of the most powerful tools in R. Many operations that normally require loops or complex programs can be solved elegantly and efficiently in R by using the power of subscripting.

Although R is usually thought of as a language for working with numbers, more and more data is appearing in the form of character strings instead of numbers. Along with basic functions for breaking apart and putting together character strings, R provides a complete implementation of regular expressions; coupled with vectorization, most character data problems can be solved simply and efficiently. Chapter 7 addresses those areas of R focused on character data.

Since most analyses, both model-based and graphical, operate on data frames, the final two chapters of the book directly address working with data frames. Chapter 8 discusses aggregation techniques, where the contents of a data frame are summarized, often broken down by groups. Chapter 9 covers the somewhat related issue of transforming and reshaping data frames. Emphasis is on methods that take advantage of R's power, and which will scale up appropriately as the size of data they operate on increases.

One aspect of this book that may seem unfamiliar is the use of the equal sign (=) as an assignment operator rather than the more traditional "gets" operator (<-). I find using the equal sign more natural than the other notation, so I've used it in all the examples. The one situation where this causes problems (assigning a value to a variable as part of a function call) is discussed in Section 8.7.

While the focus of the book is using the functions and methods that are built in to base R, a number of packages from CRAN (the Comprehensive R Archive Network) are introduced in the text. These are packages that I've personally found useful in my own work, and omission of other packages is by no means meant to imply that those packages aren't useful. In fact, with the wide variety of new packages contributed by the R community, and serious R programmer would be well advised to visit the R project homepage (http://r-project.org or preferably an appropriate mirror site) to check for new packages. Another valuable resource on this page is the R Newsletter, which often provides in-depth information on using some of the new packages.

I'd like to express my most sincere gratitude to both the original developers of the S language, the R Core development team, and the entire R community for creating such a wonderful language and inspiring its users to come up with new and exciting ways of using it.

Contents

1

Data in R

1.1 Modes and Classes

Every object in R contains a number of attributes to describe the nature of the information in that object. Two of the most important attributes of data in R are the mode and the class. When managing data, it is important to understand the differences among the different types of data that R supports, and when problems arise with data, the problem is often that the data is the incorrect mode or class for a particular operation.

The `mode` function returns the mode of any object in R, and the `class` function returns the class. When working with data, the most commonly encountered modes of individual objects are numeric, character, and logical. However, since data in R usually revolves around a collection of data (for example, a matrix or dataset), other modes will often be encountered. When deciding on how data should be stored in R, one important consideration has to do with the mode of the data being studied. Some objects (like matrices or other arrays) demand that all the data contained in them be of the same mode; others (like lists and data frames) allow for multiple modes within a single object.

In addition to the `mode` and `class` function, the `typeof` function can sometimes provide additional information about the type of an object, although it is not generally as useful as that returned by `mode` or `class`.

One other consideration when planning how data should be entered into R has to do with categorical data. R provides the factor class to store this type of data, and factors are automatically treated specially in statistical models and plotting functions. Values stored as factors require less storage than regular values, because R need only store each unique level once. If you examine the mode of a factor object, you'll notice that it is always numeric, even though it may display as character data, so special care is needed when working with factors. The `class` function, or one of the predicate functions described in Section 1.3 can be used to recognize factors once they are stored in R. Further information about factors can be found in Chapter 5.

Another important data type concerns dates and times. While this sort of information can be stored as a simple character representation, it is difficult to manipulate in this form. R provides several mechanisms for storing dates, including the built-in Date, POSIXlt, and POSIXct classes, and the contributed chron package. The differences among these different representations as well as information on manipulating dates and times are provided in Chapter 4.

Finally, one of the most often encountered modes of data is the list. Lists are the most flexible way of storing data in R, since they can accommodate objects of different modes and lengths. Many functions in R use lists to hold their results, and lists provide a very attractive way of accumulating information incrementally. When you've got a list and need to find the modes of the components of the list, the sapply function (discussed in detail in Section 8.3) can be used as shown in the following example:

```
> mylist = list(a=c(1,2,3),b=c("cat","dog","duck"),
+ d=factor("a","b","a"))
> sapply(mylist,mode)
          a           b           d
  "numeric" "character"   "numeric"
> sapply(mylist,class)
          a           b           d
  "numeric" "character"    "factor"
```

1.2 Data Storage in R

It's very rare that single values (scalars) will be the center of an R session, so one of the first questions encountered when working with data in R is what sort of object should be used to hold collections of data. The vector is the simplest way to store more than one value in R. The c function (mnemonic for catenate or combine) allows you to quickly enter data into R:

```
> x = c(1,2,5,10)
> x
[1]  1  2  5 10
> mode(x)
[1] "numeric"
> y = c(1,2,"cat",3)
> y
[1] "1"    "2"    "cat" "3"
> mode(y)
[1] "character"
> z = c(5,TRUE,3,7)
> z
[1] 5 1 3 7
> mode(z)
[1] "numeric"
```

Notice that when elements of different modes are combined with c, the mode of the resultant vector is different than that of its parts. In particular, if any of the elements are character, the other elements will be converted to characters; logical elements combined with numeric elements will be converted to numeric equivalents with TRUE becoming 1 and FALSE becoming 0. The c function can also be used to combine vectors:

```
> all = c(x,y,z)
> all
 [1] "1"    "2"    "5"    "10"   "1"    "2"    "cat" "3"    "5"
[10] "1"    "3"    "7"
```

Once again, since some of the elements of the combined vector have mode of character, the entire vector is converted to character.

The elements of the vector can be assigned names, which will be used when the object is displayed, and which can also be used to access elements of the vector through subscripts (Section 6.1). Names can be given when the vector is first created, or they can be added or changed after the fact using the names assignment function:

```
> x = c(one=1,two=2,three=3)
> x
  one   two three
    1     2     3
> x = c(1,2,3)
> x
[1] 1 2 3
> names(x) = c('one','two','three')
> x
  one   two three
    1     2     3
```

A further feature of the names assignment function is that it can be indexed to modify only selected elements of the names:

```
> names(x)[1:2] = c('uno','dos')
> x
  uno   dos three
    1     2     3
```

One surprising fact about vectors in R is that, in many cases if two vectors involved in an operation are not of the same length, R will recycle the values of the shorter vector in order to make the lengths compatible. This is a generalization of the fact that when a vector and a scalar are involved in an operation, R will silently repeat the scalar value to correspond to each value of the vector. So to add one to each element of a vector, a scalar value of 1 can be used:

```
> nums = 1:10
> nums + 1
 [1]  2  3  4  5  6  7  8  9 10 11
```

The same sort of thing will happen if the one operand is a vector of a different
length than the other:

```
> nums = 1:10
> nums + c(1,2)
 [1]  2  4  4  6  6  8  8 10 10 12
```

Note how the values 1 and 2 are repeated in order to allow the operation to
succeed. R will be silent about these kind of operations, unless the length of
the longer object is not an even multiple of the length of the shorter object:

```
> nums = 1:10
> nums + c(1,2,3)
 [1]  2  4  6  5  7  9  8 10 12 11
Warning message:
longer object length
        is not a multiple of shorter object length in:
                        nums + c(1, 2, 3)
```

Notice that this is just a warning; the operation is still carried out.

Arrays are a multidimensional extension of vectors, and, like vectors, all of
the objects of an array must be of the same mode. The most commonly used
array in R is the matrix, a 2-dimensional array. Matrices are stored internally
as vectors, with the columns of the matrix "stacked" on top of each other.
The `matrix` function converts a vector to a matrix. The `nrow=` and `ncol=`
arguments to `matrix` specify the number of rows and columns, respectively.
If only one of these arguments is given, the other will be calculated based on
the length of the input data.

Since matrices are internally stored by columns, `matrix` assumes that the
input vector will be converted to a matrix by columns; the `byrow=TRUE` argument can be used to override this in the more common case where the matrix
needs to be read in by rows. The mode of a matrix is simply the mode of
its constituent elements; the class of a matrix will be reported as `matrix`. In
addition, matrices have an attribute called `dim` which is a vector of length two
containing the number of rows and columns. The `dim` function returns this
vector; alternatively, individual elements can be accessed using the `nrow` or
`ncol` functions.

Names can be assigned to the rows and/or columns of matrices, through
the `dimnames=` argument of the `matrix` function, or after the fact through the
`dimnames` or `row.names` assignment function. Since the number of rows and
columns of a matrix need not be the same, the value of `dimnames` must be a
list; the first element is a vector of names for the rows, and the second is a
vector of names for the columns. Like vectors, these names are used for display,
and can be used to access elements of the matrix through subscripting. To

provide names for just one dimension of a matrix, use a value of NULL for the dimension for which you don't wish to provide names. For example, to create a 5×3 matrix of random numbers (See Section 2.2), and to name the columns A, B, and C, we could use statements like

```
> rmat = matrix(rnorm(15),5,3,
+                dimnames=list(NULL,c('A','B','C')))
> rmat
              A          B           C
[1,] -1.15822190 -1.1431019  0.464873841
[2,] -0.04083058  0.3705789  0.320723479
[3,] -0.25480412 -0.5972248 -0.004061773
[4,]  0.48423349 -0.8727114 -0.663439822
[5,]  1.93566841 -0.2338928 -0.605026563
```

Similarly, we could first create the matrix, then provide the dimnames separately:

```
dimnames(rmat) = list(NULL,c('A','B','C'))
```

Lists provide a way to store a variety of objects of possibly varying modes in a single R object. Note that when forming a list, the mode of each object in the list is retained:

```
> mylist = list(c(1,4,6),"dog",3,"cat",TRUE,c(9,10,11))
> mylist
[[1]]
[1] 1 4 6

[[2]]
[1] "dog"

[[3]]
[1] 3

[[4]]
[1] "cat"

[[5]]
[1] TRUE

[[6]]
[1]  9 10 11

> sapply(mylist,mode)
[1] "numeric"   "character" "numeric"   "character"
[5] "logical"   "numeric"
```

The important thing to notice about lists is that the elements of the list need
not be of the same mode; the simple example provided also shows that the
length of the elements need not be the same.

Like other objects in R, list elements can be named, either when the list
is being created, or by using the **names** assignment function if the list already
exists. The **list** function takes no keyword arguments, so list elements can
be named when they are passed to the function:

```
> mylist = list(first=c(1,3,5),second=c('one','three','five'),
+                third='end')
> mylist
$first
[1] 1 3 5

$second
[1] "one"    "three" "five"

$third
[1] "end"
```

The same result can be achieved using the **names** function after creating the
(unnamed) list:

```
> mylist = list(c(1,3,5),c('one','three','five'),'end')
> names(mylist) = c('first','second','third')
```

Many data analyses revolve around the idea of a dataset, a collection
of related values which can be treated as a single unit. For example, you
might collect information about different companies; for each company you
would have a name, an industry type, the number of employees, type of health
care plans offered, etc. For each of the companies you study you would have
values for each of these variables. If we store the data in a matrix, with rows
representing observations and columns representing variables, it would be easy
to access the data, but since the modes of the variables in a dataset will often
not be the same, a matrix would force, say, numeric variables to be stored as
character variables. To allow the ease of indexing that a matrix would provide
while accommodating different modes, R provides the data frame. A data
frame is a list with the restriction that each element of the list (the variables)
must be of the same length as every other element of the list. Thus, the mode
of a data frame is **list**, and its class is **data.frame**. While there is some
overhead for storing data in a data frame as opposed to a matrix, data frames
are the preferred method for working with "observations and variables"-style
datasets.

1.3 Testing for Modes and Classes

While the mode or class of an object can easily be examined through the mode and class functions, in many cases R provides a simpler way to verify whether an object has a particular mode, or is a member of a particular class. A large number of functions in R, beginning with the string "is.", can be used to test if an object is of a particular type. Among the many such predicate functions available in R are is.list, is.factor, is.numeric, is.data.frame, and is.character. These functions can be used to make sure that the data you're working with will behave the way that you expect, or that functions that you write will work properly with a variety of data.

Although R is not a true object-oriented language, many functions in R, collectively known as generic functions, will behave differently depending on the class of their arguments. For a given class, you can find out which functions will treat the class specially through the methods function. For more information about the object-oriented models in R, see Section 2.5.

1.4 Structure of R Objects

For simple cases such as vectors, matrices, and data frames, it's usually straightforward to determine what an object in R contains; examining the class and mode of the object, along with its length or dim attribute, should be sufficient to allow you to work effectively with the object. This process can conveniently be carried out for all the objects in a workspace with the ls.str function. However, in some cases, especially with nested lists, it can be difficult to understand how information is arranged in the object, and displaying the object in its entirety rarely elucidates the structure in these cases. The following examples are artificial, and have been kept small to reduce space, but they illustrate some strategies for getting an understanding of the structure of data in R.

Returning to an earlier example, suppose we have the following list:

```
> mylist = list(a=c(1,2,3),b=c("cat","dog","duck"),
+               d=factor("a","b","a"))
```

The summary function will provide the names, lengths, classes, and modes of the elements of the list:

```
> summary(mylist)
  Length Class  Mode
a 3      -none- numeric
b 3      -none- character
d 1      factor numeric
```

This provides useful information, but only looks at top-level elements of the list. If we have a list whose elements are lists, summary will not examine the structure of those interior lists:

```
> nestlist = list(a=list(matrix(rnorm(10),5,2),val=3),
+                        b=list(sample(letters,10),values=runif(5)),
+                        c=list(list(1:10,1:20),list(1:5,1:10)))
> summary(nestlist)
   Length Class  Mode
a 2        -none- list
b 2        -none- list
c 2        -none- list
```

In situations where direct examination provides too much detail, and summary or similar functions provide too little detail, the str function tries to provide a workable compromise. With the current example, it can be seen that str provides details about the nature of all the components of the object, presented in a display whose indentation provides visual cues to the structure of the object:

```
> str(nestlist)
List of 3
 $ a:List of 2
  ..$    : num [1:5, 1:2]  0.302 -1.534  1.133 -2.304  0.305
 ... ..$ val: num 3
 $ b:List of 2
  ..$      : chr [1:10] "v" "i" "e" "z" ...
  ..$ values: num [1:5] 0.438 0.696 0.722 0.164 0.435
 $ c:List of 2
  ..$ :List of 2
 .. ..$ : int [1:10] 1 2 3 4 5 6 7 8 9 10
 .. ..$ : int [1:20] 1 2 3 4 5 6 7 8 9 10 ...
  ..$ :List of 2
 .. ..$ : int [1:5] 1 2 3 4 5
 .. ..$ : int [1:10] 1 2 3 4 5 6 7 8 9 10
```

The number of elements displayed for each component is controlled by the vec.len= argument, and can be set to 0 to suppress any values being printed; the depth of levels displayed for each object is controlled by the max.level= argument, which defaults to NA, meaning to display whatever depth of levels is actually encountered in the object.

1.5 Conversion of Objects

To temporarily change the way an object in R behaves, a variety of conversion routines, each beginning with the string "as.", are provided. If it makes sense, these functions can be used to create an object equivalent to the one that you're working with, but which has a different mode or class. A simple example involves numbers which are stored as characters. This may occur when data is first entered into R, or it may arise as a side effect of some other operation.

Consider the `table` function, discussed in detail in Section 8.1. This function will return a vector of integer counts representing how many times each unique value in an object appears. The vector it returns is named, based on the unique values encountered. Suppose we use the table function on a vector of numbers, and then try to use this tabled version of the data to calculate a sum of all the values:

```
> nums = c(12,10,8,12,10,12,8,10,12,8)
> tt = table(nums)
> tt
nums
 8 10 12
 3  3  4
> names(tt)
[1] "8"  "10" "12"
> sum(names(tt) * tt)
Error in names(tt) * tt : non-numeric argument
        to binary operator
```

Since the error message suggests that `sum` was expecting a numeric vector, we can create a numeric version of `names(tt)` (without modifying the original version) using `as.numeric`:

```
> sum(as.numeric(names(tt)) * tt)
[1] 102
```

Of course, not all possible conversions make sense. If an inappropriate conversion is attempted, R will produce an error or warning message, and may generate missing values. (See Section 1.6.)

Note that the `as.` forms for many types of objects behave very differently than the function which bears the type's name. For example, notice the difference between the `list` function and the `as.list` function:

```
> x = c(1,2,3,4,5)
> list(x)
[[1]]
[1] 1 2 3 4 5

> as.list(x)
[[1]]
[1] 1

[[2]]
[1] 2

[[3]]
[1] 3
```

```
[[4]]
[1] 4

[[5]]
[1] 5
```

The `list` function creates a list (of length one) containing the argument it was passed while `as.list` converts the vector into a list of the same length as the vector.

One useful conversion that will take place automatically concerns logical variables. When a logical variable is used in a numeric context, each occurrence of `TRUE` will be treated as `1`, while values of `FALSE` will be treated as 0. Coupled with the vectorization of most functions, this allows many counting operations to be performed easily. For example, to find all the values in a vector, `x`, that are greater than 0, the expression `sum(x > 0)` could be used; the number of unequal elements in two matrices `a` and `b` could be calculated as `sum(a != b)`.

1.6 Missing Values

Missing values arise in data for a variety of reasons. The missing values may be part of the original data, or they may arise as part of a computation or conversion that takes place after you've read your data into R. In all cases, missing values are treated consistently, and will propagate across any computation that involves them, so it's important to recognize missing values as early as possible when you're working with data.

The value `NA`, without quotes, represents a missing value. You can assign a variable a value of `NA`, but to test for a missing value you must use the `is.na` function. This function will return `TRUE` if a value is missing and `FALSE` otherwise.

If a missing value occurs as the result of certain computations (for example, division by zero or taking the logarithm of a negative number), it may display as `Inf` or `NaN`. While the `is.na` function will recognize these values as missing, the `is.nan` function can be used to distinguish this type of missing value from the ordinary `NA` value.

1.7 Working with Missing Values

Many of the functions provided with R have arguments that are useful when your data contain missing values. Most of the statistical summary functions (`mean`, `var`, `sum`, `min`, `max`, etc.) accept an argument called `na.rm=`, which can be set to `TRUE` if you want missing values to be removed before the summary is calculated. For functions that don't provide such an argument, the negation operator (!) can be used in an expression like x[!is.na(x)] to create a vector which contains only the nonmissing values in x.

The statistical modeling functions (lm, glm, gam, etc.) all have an argument called na.action=, which allows you to specify a function that will be applied to the data frame specified by the data= argument before the modeling function processes the data. One very useful choice for this argument is na.omit, which will return a data frame with any row containing one or more missing values eliminated. Don't overlook the fact that na.omit can be called directly to create such a data frame independent of the modeling functions. The complete.cases function may also be useful to achieve the same task.

Normally, missing values are not included when a variable is made into a factor; if you want the missing values to be considered a valid factor level, use the exclude=NULL argument to factor when the factor is first created. (See Chapter 5 for more details.)

When importing data from outside sources, missing values may be represented by some string other than NA. In those cases, the na.strings= argument of read.table (Section 2.2) can be passed a vector of character values that should be treated as missing values. Since the na.strings= argument cannot be set selectively for different columns, it may sometimes be prudent to read the missing values into R in whatever form they occur, and convert them later.

2

Reading and Writing Data

2.1 Reading Vectors and Matrices

The c function has already been introduced as a way to input small amounts of data into R. When the amount of data is large, and especially when typing the data into the console is inappropriate, the scan function can be used. scan is most appropriate when all the data to be read is of the same mode, so that it can be accommodated by a vector or matrix. For reading data with variables of mixed modes, see Section 2.2.

The first argument to scan can be a quoted string or character variable containing the name of a file or a URL, or it can be any of a number of connections (Section 2.1) to allow other input sources. If no first argument is given, scan will read from the console, stopping when a completely blank line is entered.

By default, scan expects all of its input to be numeric data; this can be overridden with the what= argument, which specifies the type of data that scan will see. For example, to read a vector of character values with scan, you can specify what="":

```
> names = scan(what="")
1: joe fred bob john
5: sam sue robin
8:
Read 7 items
> names
[1] "joe"   "fred"  "bob"   "john"  "sam"   "sue"   "robin"
```

When reading from the console, R will prompt you with the index of the next item to be entered, and report on the number of elements read when it's done.

If the what= argument to scan is a list containing examples of the expected data types, scan will output a list with as many elements as there are data types provided. To specify numeric values, you can pass a value of 0:

```
> names = scan(what=list(a=0,b="",c=0))
1: 1 dog 3
2: 2 cat 5
3: 3 duck 7
4:
Read 3 records
> names
$a
[1] 1 2 3

$b
[1] "dog"   "cat"   "duck"

$c
[1] 3 5 7
```

Note that, by naming the elements in the list passed through the what= argument, the output list elements are appropriately named. When the argument to what= is a list, the multi.line= option can be set to FALSE to prevent scan from trying to use multiple lines to read the records for an observation.

One of the most common uses for scan is to read in data matrices. Since scan returns a vector, a call to scan can be embedded inside a call to the matrix function:

```
> mymat = matrix(scan(),ncol=3,byrow=TRUE)
1: 19 17 12
4: 15 18 9
7: 9 10 14
10: 7 12 15
13:
Read 12 items
> mymat
       [,1] [,2] [,3]
[1,]    19   17   12
[2,]    15   18    9
[3,]     9   10   14
[4,]     7   12   15
```

Notice the use of the byrow=TRUE argument. This allows the vector to be converted to a matrix in the way that such data is usually presented.

In order to skip fields while reading in data with scan , a type of NULL can be used in the list passed to the what= argument. Suppose we have a large data file with ten numeric fields on each line, but we only need to read the contents of the first, third, and tenth fields. We could use a call to scan as follows:

```
> values = scan(filename,
+               what=c(f1=0,NULL,f3=0,rep(list(NULL),6),f10=0))
```

Since a value of NULL will not be replicated by rep, multiple NULL values are added as lists, and the c function properly integrates them into the list passed to scan. Once the file is read in this way, a matrix with the extracted fields could be constructed with the cbind function:

```
result = cbind(values$f1,values$f3,values$f10)
```

2.2 Data Frames: `read.table`

The read.table function is used to read data into R in the form of a data frame. read.table always returns a data frame, which means that it is ideally suited to read data with mixed modes. (For data of a single mode, like numeric matrices, it is more efficient to use scan.) read.table expects each field (variable) in the input source to be separated by one or more separators, by default any of spaces, tabs, newlines or carriage returns. The sep= argument can be used to specify an alternative separator. (See Section 2.3 for convenience functions designed for comma- or tab-separated data.) If there are no consistent separators in the input data, but each variable occupies the same columns for every observation, the read.fwf function, described in Section 2.4, can be used.

If the first line in your input data consists of variable names separated by the same separator as the data, the header=TRUE argument can be passed to read.table to use these names to identify the columns of the output data frame. Alternatively, the col.names= argument to read.table can specify a character vector containing the variable names. Without other guidance, read.table will name the variables using a V followed by the column number.

The only required argument to read.table is a file name, URL, or connection object (See Section 2.1). Under Windows, make sure that double backslashes are used in pathnames, since a single backslash in a character string in R indicates that the next character should be treated specially. If your data is in the standard format as described above, that should be all that read.table needs, with the possible addition of header=TRUE. However, read.table is very flexible, and you may sometimes need to make adjustments using the features described below.

Because it offers increased efficiency in storage, read.table automatically converts character variables to factors. This may cause some problems when trying to use the variables as simple character strings. While this can usually be resolved using the methods discussed in Chapter 5, you can prevent conversion to factors by using the stringsAsFactors= argument. Passing the value FALSE through this argument will prevent any factor conversion. To insure that character variables are never converted to factors, the system option stringsAsFactors can be set to FALSE using

```
> options(stringsAsFactors=FALSE)
```

The `as.is=` argument can be used to suppress factor conversion for a subset of the variables in your data, by supplying a vector of indices specifying the columns not to be converted, or a logical vector with length equal to the number of columns to be read and `TRUE` wherever factor conversion is to be suppressed. You may notice a speedup in reading your data if you suppress some or all of the factor conversion, at the cost of increased storage space.

The `row.names=` argument can be used to pass a vector of character values to be used as row names to identify the output and which can be used instead of numeric subscripts when indexing the data frame. (See Section 6.1.) An argument of `row.names=NULL` will use a character representation of the observation number for the row names.

`read.table` will automatically treat the symbol `NA` as representing a missing value for any data type, and `NaN`, `Inf` and `-Inf` as missing for numeric data. To modify this behavior, the `na.strings` argument can be passed a vector of character values that should be interpreted as representing missing values.

By default, `read.table` will treat any text after a pound sign (#) as a comment. You can change the character used as a comment character through the `comment.char=` argument. If your input source doesn't contain any comments, setting `comment.char=''` may speed up reading your data.

For locales which use a character other than the period (.) as a decimal point, the `dec=` argument can be used to specify an alternative. The `encoding=` argument can be used to interpret non-ASCII characters in your input data.

You can control which lines are read from your input source using the `skip=` argument that specifies a number of lines to skip at the beginning of your file, and the `nrows=` argument which specifies the maximum number of rows to read. For very large inputs, specifying a value for `nrows=` which is close to but greater than the number of rows to be read may provide an increase in speed.

`read.table` expects the same number of fields on each line, and will report an error if it detects otherwise. If the unequal numbers of fields are due to the fact that some observations naturally have more variables than others, the `fill=TRUE` argument can be used to fill in observations with fewer variables using blanks or `NA`s. If `read.table` reports that there are unequal numbers of fields on some of the lines, the `count.fields` function can often help determine where the problem is.

`read.table` accepts a `colClasses=` argument, similar to the `what=` argument of `scan`, to specify the modes of the columns to be read. Since `read.table` will automatically recognize character and numeric data, this argument is most useful when you want to perform more complex conversions as the data is being read, or if you need to skip some of the fields in your input connection. Explicitly declaring the types of the columns may also improve the efficiency of reading data. To specify the column classes, provide a vector of character values representing the data types; any type for which there is an "as." method (See Section 1.3) can be used. A value of "NULL" instructs read.table to skip that column, and a value of NA (unquoted) lets read.table decide the format to use when reading that column.

2.3 Comma- and Tab-Delimited Input Files

For the common cases of reading in data whose fields are separated by commas or tabs, R provides three convenience functions, read.csv, read.csv2, and read.delim. These functions are wrappers for read.table, with appropriate arguments set for comma-, semicolon-, or tab-delimited data, respectively. Since these functions will accept any of the optional arguments to read.table, they are often more convenient than using read.table and setting the appropriate arguments manually.

2.4 Fixed-Width Input Files

Although not as common as white-space-, tab-, or comma-separated data, sometimes input data is stored with no delimiters between the values, but with each variable occupying the same columns on each line of input. In cases like this, the read.fwf function can be used. The widths= argument can be a vector containing the widths of the fields to be read, using negative numbers to indicate columns to be skipped. If the data for each observation occupies more than one line, widths= can be a list of as many vectors as there are lines per observation. The header=, row.names=, and col.names= arguments behave similarly to those in read.table.

To illustrate the use of read.fwf, consider the following lines, showing the 10 counties of the United States with the highest population density (measured in population per square mile):

```
New York, NY              66,834.6
Kings, NY                 34,722.9
Bronx, NY                 31,729.8
Queens, NY                20,453.0
San Francisco, CA         16,526.2
Hudson, NJ                12,956.9
Suffolk, MA               11,691.6
Philadelphia, PA          11,241.1
Washington, DC             9,378.0
Alexandria IC, VA          8,552.2
```

Since the county names contain blanks and are not surrounded by quotes, read.table will have difficulty reading the data. However, since the names are always in the same columns, we can use read.fwf. The commas in the population values will force read.fwf to treat them as character values, and, like read.table, it will convert them to factors, which may prove inconvenient later. If we wanted to extract the state values from the county names, we might want to suppress factor conversion for these values as well, and as.is=TRUE will be used. Assuming that the data is stored in a file named city.txt, the values could be read as follows:

```
> city = read.fwf("city.txt",widths=c(18,-19,8),as.is=TRUE)
> city
                 V1          V2
1  New York, NY        66,834.6
2  Kings, NY           34,722.9
3  Bronx, NY           31,729.8
4  Queens, NY          20,453.0
5  San Francisco, CA   16,526.2
6  Hudson, NJ          12,956.9
7  Suffolk, MA         11,691.6
8  Philadelphia, PA    11,241.1
9  Washington, DC       9,378.0
10 Alexandria IC, VA    8,552.2
```

Before using V2 as a numeric variable, the commas would need to be removed using gsub (see Section 7.8):

```
> city$V2 = as.numeric(gsub(',','',city$V2))
```

2.5 Extracting Data from R Objects

While previous sections have discussed working with data stored in built-in classes, R provides two mechanisms for developers to define their own classes, so it's important to understand how data is stored in such objects. The class mechanisms in R provide some of the features of object-oriented programming, namely, method dispatch and inheritance. Method dispatch allows R to examine the class of the arguments to a function, and to invoke a special version of the function designed for that class of object. Not all functions in R provide method dispatch; the ones that do are known as generic functions. Inheritance allows developers to create new classes that are similar to other classes; only methods that differ from the original class need to be provided. When an object in R inherits the properties of an already defined object, its class attribute will be a vector containing the object's class (in the first position), along with the classes from which it inherits.

In the first mechanism for object orientation in R, known as S3 or "old-style" classes, method dispatch is implemented for generic functions by the existence of a function whose name is of the form function.class. If no such function is found in the search path, a function whose name is of the form function.default will be invoked, and default functions exist for all the S3 generics. S3 generic functions can be recognized because their body consists of a call to the UseMethod function, which actually performs the dispatch. It's important to recognize when generic functions are being called, because help pages for specific method/object combinations may be available through their "full" names. For example, the help page for the summary function doesn't discuss any properties of the method that will be invoked when summary is passed an lm object; the help page for summary.lm would have to be accessed directly. Even though you may refer to these specific methods to view their help pages, it's rarely if ever necessary to call them directly—they should always be called through the generic function.

As an example, consider the lm function, which performs linear model calculations. A object returned by this function will have class of lm; when the object is printed or displayed, R will look for a function called print.lm, which will display appropriate information about the linear model which was fit. For example, the following code produces an lm object, and then displays it through the print function:

```
> slm = lm(stack.loss ~ Air.Flow + Water.Temp,data=stackloss)
> class(slm)
[1] "lm"
> slm

Call:
lm(formula = stack.loss ~ Air.Flow + Water.Temp,
    data = stackloss)

Coefficients:
(Intercept)      Air.Flow     Water.Temp
   -50.3588        0.6712         1.2954
```

When a class is created, a set of functions to extract data from objects of that class is usually also provided. These so-called accessor functions are the recommended way to extract information from objects in R, since they will provide a stable interface even if the internal structure of the object changes. Because of the naming convention used in S3 method dispatch, the apropos function can be used to find all the available methods for a given class:

```
> apropos('.*\\.lm$')
 [1] "anovalist.lm"    "anova.lm"         "hatvalues.lm"
 [4] "model.frame.lm"  "model.matrix.lm" "plot.lm"
 [7] "predict.lm"      "print.lm"         "residuals.lm"
[10] "rstandard.lm"    "rstudent.lm"      "summary.lm"
[13] "kappa.lm"
```

(If any functions are marked nonvisible, the getAnywhere function can be used to see them.) Thus, to get the predicted values from an lm object, the predict function will dispatch to predict.lm. It's worth repeating that you should avoid calling a function like predict.lm directly in favor of relying on the generic function (in this case, `predict`). Also note that if an object has multiple classes, you should look for relevant functions designed for any of the classes from which the object inherits.

Most S3 objects are stored as lists, so if an appropriate accessor function is not available, data can be extracted directly from the object by treating it as a list. The first step in these cases is to use the `names` function to find the available elements:

```
> names(slm)
 [1] "coefficients"   "residuals"       "effects"
 [4] "rank"           "fitted.values"  "assign"
 [7] "qr"             "df.residual"    "xlevels"
[10] "call"           "terms"          "model"
```

Now, for example, we could find the residual degrees of freedom for the model by extracting it directly:

```
> slm$df.residual
[1] 18
```

or

```
> slm['df.residual']
[1] 18
```

Since the method dispatch provided by the old-style classes is limited to using only the first argument to the function, and since the naming conventions can sometimes lead to confusion, a more formal method of defining classes (known as "new-style" or S4 classes) has also been developed. This is the preferred way to implement new classes in R, and will become much more prevalent over time. Some of the functions required to work with new-style classes are found in the `methods` package, so if they are not available, they can be loaded using

```
> library(methods)
```

With S4 classes, generic functions can be identified by the presence of a call to the `standardGeneric` function inside the generic function's definition.

As an example of an S4 class, consider the `mle` function, used for maximum likelihood estimation, and found in the `stats4` package. We'll simulate data from a gamma distribution, and then use `mle` to estimate the parameters for that distribution:

```
> library(stats4)
> set.seed(19)
> gamdata = rgamma(100,shape=1.5,rate=5)
> loglik = function(shape=1.5,rate=5)
+           -sum(dgamma(gamdata,shape=shape,rate=rate,log=TRUE))
> mgam = mle(loglik)
```

The isS4 function can be used to determine whether or not an object is using the old-style or new-style classes:

```
> class(mgam)
[1] "mle"
attr(,"package")
[1] "stats4"
> isS4(mgam)
[1] TRUE
```

As with old-style methods the first choice for accessing information from an S4 class should always be using the accessor functions provided along with the function that created the object. For S4 classes, it's easy to find the available methods using the showMethods function (from the methods package):

```
> showMethods(class='mle')
Function: coef (package stats)
object="mle"

Function: confint (package stats)
object="mle"

Function: initialize (package methods)
.Object="mle"
    (inherited from: .Object="ANY")

Function: logLik (package stats)
object="mle"

Function: profile (package stats)
fitted="mle"

Function: show (package methods)
object="mle"

Function: summary (package base)
object="mle"

Function: update (package stats)
object="mle"

Function: vcov (package stats)
object="mle"
```

Thus, for example, the variance–covariance matrix of the estimators is available by using the vcov function:

```
> vcov(mgam)
           shape        rate
shape 0.05464054 0.1724472
rate  0.17244719 0.7228044
```

Naturally, the help page for the function that produced the object, as well as additional help pages describing the class (if available), can be consulted for additional information.

Although there is no generic `print` function for S4 classes, the generic `show` function takes its place for printing or displaying S4 classes.

The actual entities that compose an S4 object are stored in so-called slots. To see the available slots in an object, the `showClass` function can be used. If it becomes necessary to access the slots directly, the `@` operator can be used in a fashion similar to the `$` operator. Following the `mle` example, suppose we wanted to retrieve the function that was used to calculate the likelihood which was stored in `mgam`:

```
> getClass(class(mgam))
Slots:

Name:       call     coef  fullcoef       vcov        min
Class:  language  numeric   numeric     matrix    numeric

Name:     details minuslogl     method
Class:       list  function  character
> mgam@minuslogl
function(shape=1.5,rate=5)
    -sum(dgamma(gamdata,shape=shape,rate=rate,log=TRUE))
```

The `slot` function can be used if the name of the desired slot is stored in a character variable:

```
> want = 'minuslogl'
> slot(mgam,want)
function(shape=1.5,rate=5)
    -sum(dgamma(gamdata,shape=shape,rate=rate,log=TRUE))
```

For both styles of classes, some of the methods provided may create objects containing additional information about the objects they operate on. This is especially true for the `summary` method for many objects. Returning to the `lm` example, we can examine what's available through the `summary` method by creating a `summary` object from the `lm` object, and examining its names:

```
> sslm = summary(slm)
> class(sslm)
[1] "summary.lm"
```

```
> names(sslm)
 [1] "call"          "terms"         "residuals"
 [4] "coefficients"  "aliased"       "sigma"
 [7] "df"            "r.squared"     "adj.r.squared"
[10] "fstatistic"    "cov.unscaled"
```

As can be seen, a number of useful quantities have been calculated through the summary method, and are available through their named components in the resulting summary.lm object.

2.6 Connections

Connections provide a flexible way for R to read data from a variety of sources, providing more complete control over the nature of the connection than simply specifying a file name as input to functions like read.table and scan. Table 2.1 lists some of the functions in R which can create a connection.

Function	Data source
file	Files on the local file system
pipe	Output from a command
textConnection	Treats text as a file
gzfile	Local gzipped file
unz	Local zip archive (with single file;read-only)
bzfile	Local bzipped file
url	Remote file read via http
socketConnection	socket for client/server programs

Table 2.1. Connections

When you create a connection object, it simply defines the object; it does not automatically open the object. If a function which accepts connections receives a connection which has not been opened, it will always open it, then close it at the end of the function invocation. Thus, in the usual case, you can simply pass the connection to the function that will operate on it, without worrying about when the connection will be opened or closed. If a connection doesn't behave the way you expect, or if you're not sure if you've already closed it, the isOpen function can be used to test if a connection is open; an optional second argument, set equal to "read" or "write" can test the mode with which it was opened.

One exception to this scheme is the case where a file is read in pieces, for example through the readLines function. If an (unopened) connection is passed to this function inside a loop, it will repeatedly open and close the file each time it's called, reading the same data over and over. To take more control over the connection, it can either be passed to the open function, or an optional mode can be provided as the second argument to the function that created the connection. Note that in this case, the connection will not be automatically closed; you must explicitly pass the connection to the `close` function.

As an illustration of this technique, consider the R project homepage, `http://www.r-project.org/main.shtml`. The latest version number of R is displayed on that page, followed by the phrase "has been released". The following program opens the connection to this URL by passing a mode of "r" for read, then reads each line until it finds the one with the latest version number:

```
> rpage = url('http://www.r-project.org/main.shtml','r')
> while(1){
+      l = readLines(rpage,1)
+      if(length(l) == 0)break;
+      if(regexpr('has been released',l) > -1){
+              ver = sub('</a.*$','',l)
+              print(gsub('^ *','',ver))
+              break
+      }
+ }
[1] "R version 2.2.1"
> close(rpage)
```

The second argument to `readLines` specifies the number of lines to read, where a value of -1 means to read everything that the connection provides. Although it may seem inefficient to read only one line at a time, the actual reads are being performed by the operating system and are buffered in memory, so that you can choose whatever number of lines is most convenient. By using this technique, we only need to process as much of the connection as necessary.

Note that connections can be used anywhere a file name could be passed to functions like `scan`, `read.table`, `write.table`, and `cat`. So to write a gzipped, comma-separated version of a data frame, we could use:

```
gfile = gzfile('mydata.gz')
write.table(mydata,sep=',',file=gfile)
```

The `write.table` function takes care of opening and closing the gzipped file, since it was not explicitly opened.

A `textConnection` can often be useful when you need to test a function that only operates on files. For example, in Section 2.2, the `colClasses` argument was introduced as a way to automatically convert data into appropriate R objects. Suppose we want to test conversion to `Date` objects (Section 4.1) using this argument. First, we create a `textConnection` with the kind of data we'll be using:

```
> sample = textConnection('2000-2-29 1 0
+ 2002-4-29 1 5
+ 2004-10-4 2 0')
```

Now we can use `sample` in place of a file name in any function that's expecting a file name:

```
> read.table(sample,colClasses=c('Date',NA,NA))
          V1 V2 V3
1 2000-02-29  1  0
2 2002-04-29  1  5
3 2004-10-04  2  0
```

The `unz` function allows read-only access to zipfiles. Since a zipfile is an archive, potentially containing many files, an additional argument to `unz` is required to specify which file you wish to extract. For example, suppose we have a zip file called `data.zip` containing several files, and we wish to create a connection to read the file called `mydata.txt` into a vector with the `scan` function. The following code could be used:

```
mydata = scan(unz('data.zip','mydata.txt'))
```

Only one file can be extracted at a time with `unz`.

When you need to explicitly open a connection (either by passing a mode to the function that creates the connection, or calling `open` directly), you can specify any of the following modes: `"w"` for write, `"r"` for read, or `"a"` for append. You can append a `t` to the end of any of these modes to specify a text connection, or a `b` to specify a binary connection. While there is no distinction between text and binary on UNIX systems, on Windows it is required to include a `b` anytime you want to operate on a file that has any nonprintable characters. In addition, specifying a binary file for writing on Windows will cause R to use UNIX-style line endings (a single newline) instead of Windows-style line endings (newline plus carriage return, sometimes displayed as control-M on UNIX systems). For more complex situations, such as opening a file for both reading and writing, consult the help file for `open`.

2.7 Reading Large Data Files

Since `readLines` and `scan` don't need to read an entire file into memory, there are situations where very large files can be processed by R in pieces. For example, suppose we have a large file containing numeric variables, and we wish to read a random sample of that file into R. Of course, if we could accommodate the entire dataset in memory, a call to `sample` (Section 2.9.1) could extract such a sample, but we'll assume that the file in question is too large to read into R in its entirety. The strategy is to select a random sample of rows before reading the data, and then extracting the selected rows as the dataset is being read in pieces. To avoid memory allocation problems, the entire matrix is preallocated before the reading begins. These ideas are implemented in the following function:

```
readbig = function(file,samplesz,chunksz,nrec=0){
    if(nrec <= 0)nrec = length(count.fields(file))
    f = file(file,'r')
    on.exit(close(f))
    use = sort(sample(nrec,samplesz))
    now = readLines(f,1)
    k = length(strsplit(now,' +')[[1]])
    seek(f,0)

    result = matrix(0,samplesz,k)

    read = 0
    left = nrec
    got = 1
    while(left > 0){
        now = matrix(scan(f,n=chunksz*k),ncol=k,byrow=TRUE)
        begin = read + 1
        end = read + chunksz
        want = (begin:end)[begin:end %in% use] - read
        if(length(want) > 0){
            nowdat = now[want,]
            newgot =  got + length(want) - 1
            result[got:newgot,] = nowdat
            got = newgot + 1
        }
        read = read + chunksz
        left = left - chunksz
    }
    return(result)
}
```

If the number of records in the file, nrec, is specified as zero or a negative number, the function calculates the number of lines in the file through a call to count.fields; your operating system may provide a more efficient way of achieving this, such as the wc -l command on Linux or Mac OS X, or the find /c command on Windows, searching for the separator character which will be present on each line. Suppose we have a comma-separated file called comma.txt in the current directory. Under Windows, we could calculate the number of lines in the file using

```
> nrec = as.numeric(shell('type "comma.txt" | find /c ","',
+                                     intern=TRUE))
```

while under UNIX-like systems, the command would be

```
> nrec = as.numeric(system('cat comma.txt | wc -l',
+                             intern=TRUE))
```

To calculate the number of columns in the file, a single line is read using readLines, and the strsplit function is called with an appropriate separator, in this case one or more blanks. Then the file is repositioned to its origin with the seek command, to prepare to actually read the data.

For optimum results, the chunksz argument can be adjusted for a given situation, but most reasonable values will result in acceptable performance.

2.8 Generating Data

Even though one of the main motivations in learning or working with R is to analyze existing data, sometimes it may be advantageous to create a data set from within R. This would be necessary, for example, to carry out a simulation, but may also be useful if you want to test a new technique or determine if a program may be appropriate for a very large dataset. In the following subsections, we'll look at a number of ways in R to generate vectors of data which can be used for simulations or testing programs when "real" data isn't available.

2.8.1 Sequences

To generate a sequence of integers between two values, the colon operator (:) can be used. For example, to create a vector of the integers from 1 to 10, we can use

```
> 1:10
 [1]  1  2  3  4  5  6  7  8  9 10
```

To get more control over the sequence, the seq function can be used. This function allows an optional increment through the by= argument, as well as more options for determining the length of the output sequence. In its simplest form, it behaves like the colon operator:

```
> seq(1,10)
 [1]  1  2  3  4  5  6  7  8  9 10
```

To create a vector of values from 10 to 100, each element separated by 5, we could use:

```
> seq(10,100,5)
 [1]  10  15  20  25  30  35  40  45  50  55  60  65  70  75
[15]  80  85  90  95 100
```

Alternatively, we can specify the length of the generated sequence instead of providing the end point:

```
> seq(10,by=5,length=10)
 [1] 10 15 20 25 30 35 40 45 50 55
```

One common use of sequences is to generate factors corresponding to the levels of a designed experiment. Suppose we wish to simulate the levels for an experiment with three groups and five subgroups, with two observations in each subgroup, for a total of 30 observations. The gl function (mnemonic for "generate levels") has two required arguments. The first is the number of different levels desired, and the second is the number of times each level needs to be repeated. An optional third argument specifies the length of the output vector. To generate a data frame composed of vectors representing the three groups, five subgroups, and two observations, we could use gl as follows:

```
> thelevels = data.frame(group=gl(3,10,length=30),
+                          subgroup=gl(5,2,length=30),
+                          obs=gl(2,1,length=30))
> head(thelevels)
  group subgroup obs
1     1        1   1
2     1        1   2
3     1        2   1
4     1        2   2
5     1        3   1
6     1        3   2
```

Further control over the output of gl can be obtained with the optional labels= argument. gl also accepts an ordered=TRUE argument to produce ordered factors.

To create a data frame with the unique combinations defined by a collection of sequences, the expand.grid function can be used. This function accepts any number of sequences, and returns a data frame with one row for each unique combination of the values passed as input. Alternatively, all of the vectors can be passed to expand.grid as a single list. Suppose we wanted to create a data frame with one observation for each combination of odd and even integers between 1 and 5. We could use expand.grid as follows:

```
> oe = expand.grid(odd=seq(1,5,by=2),even=seq(2,5,by=2))
> oe
  odd even
1   1    2
2   3    2
3   5    2
4   1    4
5   3    4
6   5    4
```

Note that the column generated from the first argument varies the most quickly, and that the sequences passed to expand.grid need not be of the same length. The number of rows in the output data frame will always be equal to the product of the length of all the sequences passed to expand.grid.

One important use of data frames produced by expand.grid is to evaluate a function over a range of parameter values. The apply function, discussed in Section 8.4, is the most effective tool for this purpose. For example, suppose we wanted to evaluate the function $x^2 + y^2$ for various values of x and y in the range of 0 to 10. First, we can generate a matrix of input values using expand.grid:

```
> input = expand.grid(x=0:10,y=0:10)
```

Now we can use apply to calculate the function for each row of the data frame returned by expand.grid, and use cbind (Section 9.6) to combine it to the input data:

```
> res = apply(input,1,function(row)row[1]^2 + row[2]^2)
> head(cbind(input,res))
  x y res
1 0 0   0
2 1 0   1
3 2 0   4
4 3 0   9
5 4 0  16
6 5 0  25
```

2.8.2 Random Numbers

Function	Distribution	Function	Distribution
rbeta	Beta	rlogis	Logistic
rbinom	Binomial	rmultinom	Multinomial
rcauchy	Cauchy	rnbinom	Negative Binomial
rchisq	Chi-square	rnorm	Normal
rexp	Exponential	rpois	Poisson
rf	F	rsignrank	Signed Rank
rgamma	Gamma	rt	Student's t
rgeom	Geometric	runif	Uniform
rhyper	Hypergeometric	rweibull	Weibull
rlnorm	Log Normal	rwilcox	Wilcoxon Rank Sum

Table 2.2. Random number generators

If you are creating a dataset for a simulation, or to test a function in R for which there is no real data available, R provides a large number of random number generators, listed in Table 2.2. The first argument to all of the random number generation functions is the number of random numbers desired; additional arguments allow specification of the parameters of the underlying distribution, and will vary depending on which distribution you are working with. Consult the help file for the function for further details.

The state of the random number generator that underlies all of the functions in Table 2.2 is stored in an object called .Random.seed. To create a reproducible sequence, an integer can be passed to the set.seed function insuring that an identical stream of random numbers will be generated whenever set.seed is set to the same value.

2.9 Permutations

2.9.1 Random Permutations

The sample function conveniently provides random permutations of either a vector of values (if the first argument is a vector), or of indices starting from one (if the first argument is a single number). With only one argument, sample returns a vector of length equal to the number of elements (if the argument is a vector) or the value of the argument if it is a number, sampling without replacement, so that each element of the input will appear exactly once in the output. Two optional arguments can override these defaults; the size= argument will return a vector of the designated size, and the replace= argument, if set to TRUE, will allow the possibility of elements of the input to appear more than once in a given sample. If the size of the desired sample is greater than the number of elements implied by the first argument, the replace= must be set to TRUE. Finally, if some elements of the input should be sampled at a higher probability than others, the optional prob= argument can provide a vector of sampling probabilities.

2.9.2 Enumerating All Permutations

Since the sample function provides a random permutation of its input, it will not be very effective in generating all possible permutations for a sequence, since it is possible that some permutations will appear more often than others. In cases like this, the permn function of the combinat package (available from CRAN) can be used. Like sample, the first argument to permn is either a vector or a single number. Called with a single argument, it returns a list containing each possible permutation of the input sequence. The optional fun= argument can be used to specify a function which will be applied to each permutation in the output list. Since the number of permutations of n elements is $n!$ (n factorial), the size of the output from permn can be very large, even for moderate values of n. The factorial function (or the fact function of the `combinat` package) can be used to calculate how many permutations exist.

If the number of permutations generated is too large to be accommodated in memory, the `numperm` function of the `sna` package(available from CRAN) can be used. This function accepts two arguments: the first represents the length of the sequence, and the second represents the specific permutation desired. Thus, if the entire set of permutations is too large to be held in memory, the `numperm` function can be used inside a loop which successively increments an index (the second argument to `numperm`) from 1 to `factorial(n)`, where n is the length of the sequence.

2.10 Working with Sequences

R provides several functions which are useful when working with sequences of numbers. The `table` function, introduced in Section 8.1, can tabulate the number of occurrences of each value in a sequence. To get just the unique values in a sequence, the `unique` function can be used. Alternatively, the `duplicated` function can be used to return a vector of logical values indicating whether each value in the sequence is duplicated; `!duplicated(x)` will return a logical vector which will be true for the unique values. In both cases, the results will be in the order that the values are encountered in the vector being studied.

The `rle` (run-length encoding) function can be used to solve a variety of problems regarding consecutive identical values in a sequence. The returned value from `rle` is a list with two components: `values`, a vector which contains the repeated values that were found, and `lengths`, a vector of the same length as `values` which tells how many consecutive values were observed. If there are no repeated values, all of the elements of `lengths` will be 1:

```
> sequence = sample(1:10)
> rle(sequence)
Run Length Encoding
  lengths: int [1:10] 1 1 1 1 1 1 1 1 1 1
  values : int [1:10] 10 5 2 8 3 1 7 4 6 9
```

As an example of the use of `rle`, suppose we have a sequence of integers, and we wish to know if there are 3 or more consecutive appearances of the value 2. Considering the return value of `rle`, this will mean that 2 will appear in the `values` component returned by `rle` with a corresponding entry in the `lengths` component with a value of three or more:

```
> seq1 = c(1,3,5,2,4,2,2,2,7,6)
> rle.seq1 = rle(seq1)
> any(rle.seq1$values == 2 & rle.seq1$lengths >= 3)
[1] TRUE
```

```
> seq2 = c(7,5,3,2,1,2,2,3,5,8)
> rle.seq2 = rle(seq2)
> any(rle.seq2$values == 2 & rle.seq2$lengths >= 3)
[1] FALSE
```

To find the location within a sequence where a particular combination of values and lengths occurs, the cumsum function can be applied to the lengths component of the returned value of rle. The returned value from cumsum will provide the index where each set of repeated values terminates; thus by using the which function as an index into this vector, we can find the end points of any desired run of values.

Continuing with the previous example, we can find the index into seq1 where a sequence of more than three values of 2 terminates:

```
> seq1 = c(1,3,5,2,4,2,2,2,7,6)
> rle.seq1 = rle(seq1)
> index = which(rle.seq1$values == 2 & rle.seq1$lengths >= 3)
> cumsum(rle.seq1$lengths)[index]
[1] 8
```

This indicates that the run of three or more values of 2 ended at position 8. To find the index where a run began, we can adjust the subscript used to cumsum, taking special care to properly handle runs at the beginning of a sequence:

```
> index = which(rle.seq1$values == 2 & rle.seq1$lengths >= 3)
> newindex = ifelse(index > 1,index - 1,0)
> starts = cumsum(rle.seq1$lengths)[newindex] + 1
> if(0 %in% newindex)starts = c(1,starts)
> starts
```

Due to the vectorization of these operators, multiple runs can be accommodated by the same strategy:

```
> seq3 = c(2,2,2,2,3,5,2,7,8,2,2,2,4,5,9,2,2,2)
> rle.seq3 = rle(seq3)
> cumsum.seq3 = cumsum(rle.seq3$lengths)
> myruns = which(rle.seq3$values == 2 &
+                 rle.seq3$lengths >= 3)
> ends = cumsum.seq3[myruns]
> newindex = ifelse(myruns > 1,myruns - 1,0)
> starts = cumsum.seq3[newindex] + 1
> if(0 %in% newindex)starts = c(1,starts)
> starts
[1]  1 10 16
> ends
[1]  4 12 18
```

For more complex situations, a logical expression can often be used as an argument to `rle`. For example, to find the location of five or more successive values greater than zero in a sequence of random numbers, we can use the following approach:

```
> set.seed(19)
> randvals = rnorm(100)
> rle.randvals = rle(randvals > 0)
> myruns = which(rle.randvals$values == TRUE &
+                  rle.randvals$lengths >= 5)
> any(myruns)
[1] TRUE
> cumsum.randvals = cumsum(rle.randvals$lengths)
> ends = cumsum.randvals[myruns]
> newindex = ifelse(myruns > 1,myruns - 1,0)
> starts = cumsum.randvals[newindex] + 1
> if(0 %in% newindex)starts = c(1,starts)
> starts
[1] 47
> ends
[1] 51
> randvals[starts:ends]
[1] 0.5783932 0.8276480 1.3111752 0.1783597 1.7036697
```

2.11 Spreadsheets

Spreadsheets, especially Microsoft Excel spreadsheets, are one of the most common methods of distributing data, and R provides several ways to access them. The simplest method, which may also be the most flexible, is to use a spreadsheet program to write the data to a comma- or tab-separated file, and then use the methods described in Section 2.2. This method is especially useful when there is additional nondata material (like headings and notes) in the spreadsheet, since such material can be removed by editing the file derived from the spreadsheet.

There are some situations, however, where this may not be feasible. An updated spreadsheet may need to be accessed every day, or an analysis may require data that comes from several spreadsheets, or from several sheets within a single spreadsheet. The next sections will look at ways of accessing spreadsheets through R functions, without the need to dump them to files.

2.11.1 The RODBC Package on Windows

On the Windows platform, spreadsheets can be read directly using the `ODBConnectExcel` function from the `RODBC` package, available from CRAN.

(For information about using RODBC to read databases, see Chapter 3.)
This function provides an interface to Excel spreadsheets using the SQL language familiar to databases, and does not require Excel itself to be installed on your computer.

The RODBC interface treats the various sheets stored in a spreadsheet file as database tables. To use the interface, a connection object is obtained by providing the pathname of the Excel spreadsheet file to odbcConnectExcel. For example, suppose a spreadsheet is stored in the file
c:\Documents and Settings\user\My Documents\sheet.xls. To get a connection object, we could use the following call:

```
> library(RODBC)
> sheet = 'c:\\Documents and Settings\\user\\My Documents
            \\sheet.xls'
> con = odbcConnectExcel(sheet)
```

Note the use of double slashes in the file name; this is used because the backslash has special meaning in R character strings, namely to inform R that certain characters need to be treated specially. Often the first step in working with spreadsheets in this way is to look at the names of the available sheets. This can be done with the sqlTables command. Continuing with the current example, we could find the names of the sheets in the sheet.xls spreadsheet by issuing the command

```
> tbls = sqlTables(con)
```

and then examining tbls$TABLE_NAME, the column of the returned data frame that contains the sheet names. From this point on, each of the sheets can be treated like a separate database table. Thus, to extract the contents of the first sheet of the database to a data frame called data1, we could use the following commands:

```
> qry = paste("SELECT * FROM",tbls$TABLE_NAME[1],sep=' ')
> result = sqlQuery(con,qry)
```

If the table name contains special characters, like spaces, brackets, or dollar signs, then it needs to be surrounded by backquotes ('). As a precautionary measure, it may be advisable to include the backquotes in all queries:

```
> qry = paste("SELECT * FROM '",tbls$TABLE_NAME[1],"'",sep="")
> result = sqlQuery(con,qry)
```

Most SQL queries will work using this method.

2.11.2 The gdata Package (All Platforms)

An alternative to using the RODBC package is the read.xls function of the gdata package, available from CRAN. This function uses a module developed for the scripting language perl (http://perl.org), and thus requires perl to be installed on your computer. This will be the case for virtually all Mac OS

X, Unix, and Linux computers, but on Windows an installation of perl will
be necessary. (The perl installer and instructions can be found on the above-
referenced web site.) `read.xls` translates a specified sheet of a spreadsheet
to a comma-separated file, and then calls `read.csv` (see Section 2.3). Thus,
any option accepted by `read.csv` can be used with `read.xls`. The `skip=` and
`header=` arguments are especially useful to avoid misinterpreting headers and
notes as data, and the `as.is=TRUE` argument can be used to suppress factor
conversion.

2.12 Saving and Loading R Data Objects

In situations where a good deal of processing must be used on a raw dataset
in order to prepare it for analysis, it may be prudent to save the R objects
you create in their internal binary form. One attractive feature of this scheme
is that the objects created can be read by R programs running on different
computer architectures than the one on which they were created, making it
very easy to move your data between different computers. Each time an R
session is completed, you are prompted to save the workspace image, which is
a binary file called `.RData` in the working directory. Whenever R encounters
such a file in the working directory at the beginning of a session, it automatically loads it making all your saved objects available again. So one method
for saving your work is to always save your workspace image at the end of
an R session. If you'd like to save your workspace image at some other time
during your R session, you can use the `save.image` function, which, when
called with no arguments, will also save the current workspace to a file called
`.RData` in the working directory.

Sometimes it is desirable to save a subset of your workspace instead of the
entire workspace. One option is to use the `rm` function to remove unwanted
objects right before exiting your R session; another possibility is to use the
`save` function. The `save` function accepts multiple arguments to specify the
objects you wish to save, or, alternatively, a character vector with the names
of the objects can be passed to `save` through the `list=` argument. Once the
objects to be saved are specified, the only other required option is the `file=`
option, specifying the destination of the saved R object. Although there is no
requirement to do so, it is common to use a suffix of `.rda` or `.RData` for saved
R workspace files.

For example, to save the R objects x, y, and z to a file called `mydata.rda`,
the following statements could be used:

```
> save(x,y,z,file='mydata.rda')
```

If the names of the objects to be saved are stored as character vectors (for
example, from the output of the `objects` function), the `list=` argument can
be used:

```
> save(list=c('x','y','z'),file='mydata.rda')
```

Once the data is saved, it can be reloaded into a running R session with the load command, whose only required argument is the name of the file to be loaded. For example, to load the objects contained in the mydata.rda file, we can use the following command:

```
> load('mydata.rda')
```

The load command can also be used to load workspaces stored in .RData files in other directories by specifying their complete file path.

2.13 Working with Binary Files

While the natural way to store R objects is through the save command, other programs may also produce binary files which, not surprisingly, use their own format and are not readable by load. The readBin and writeBin functions provide a flexible way to read and write such files. It should be noted that a fairly complete knowledge of the format of a non-R binary file is required before readBin will be able to read it. However, for well-documented file formats, readBin should be able to access the full information contained in the file. Each call to readBin will read as many values as required, but a single call can only read one type of data. The types of data that readBin can understand include double precision numeric data, integers, character strings (although the readChar and writeChar functions provide additional flexibility), complex numbers, and raw data. Since multiple calls to readBin will have to be used if there is a mixture of data types in the file, it is usually necessary to pass a connection object to readBin, so that it will not automatically re-open the file each time it is called.

As an example of the use of readBin, consider a binary file called data.bin, which consists of 20 records, each containing one integer followed by five double-precision values. Such a file could be produced, for example, using the low-level write function in a C program. The first step is opening a file connection:

```
> bincon = file('data.bin','rb')
```

Note that R will not allow access to files through readBin or writeBin unless the file is opened in binary mode; thus the value of 'rb' is passed to file to specify the mode. (The "r" stands for read, and the "b" stands for binary.)

For efficiency's sake, it's a good idea to preallocate memory for the vector or matrix which will hold the output from readBin. In this case, we can use a 20×6 matrix, and store the integer and five doubles in the rows of the matrix:

```
> result = matrix(0,20,6)
> for(i in 1:20){
+    theint = readBin(bincon,integer(),1)
+    thedoubles = readBin(bincon,double(),5)
+    result[i,] = c(theint,thedoubles)
+ }
> close(bincon)
```

As always when opening a connection inside of R, it's a good idea to close the file when you're done.

While no problems should be encountered if the data to be read was written on a computer with the same architecture as the one on which it is to be read, problems will sometimes occur if binary data from other architectures is used. There are two ways of storing data on a computer, depending on the order in which the bits of a binary value are stored; these two types are known as "little-endian" and "big-endian". Among some of the common architectures, x86 and its derivatives are little-endian, while the PowerPC and SPARC platforms are big-endian. readBin and writeBin each accept an endian= argument, which take values of "big", "little", or "swap". (Note that endianness is not an issue when using the save and load commands since R uses the same format on all architectures for its saved objects.)

Writing binary files is essentially the reverse of reading them. writeBin can only write vectors of character, numeric, logical, or complex values; in particular lists or factors will need to be converted before writing.

As an example of using writeBin, consider a data frame constructed from the state.x77 matrix:

```
> mystates = data.frame(name=row.names(state.x77),state.x77,
+                       row.names=NULL,stringsAsFactors=FALSE)
```

Note that the stringsAsFactors=FALSE argument was used to avoid factor conversion which could cause writeBin to fail. For existing datasets, the as.character function could be used to convert factors to character variables.

Suppose we wish to write a binary version of each row of the mystate data frame to a file. When writeBin converts a character variable, it uses the C programming language convention of terminating the string with a binary zero. If the program to be reading the data requires fixed-width fields, the sprintf function can be used to convert variable-length character values to fixed length. For example, to make all the elements in mystate$name the same length we can use the sprintf function as follows:

```
> maxl = max(nchar(mystates$name))
> mystates$newname = sprintf(paste('%-',maxl,'s',sep=''),
+                            mystates$name)
```

Omitting the minus sign (-) would pad the strings at the beginning, instead of the end.

Since R knows the size and types of its own objects, there is no need to explicitly provide this information to `writeBin`, but if you want `writeBin` to use a nonnative size for any of its output, the `size=` argument is available. Note that using nonnative sizes may make it difficult or impossible to read a binary file on other architectures.

We can now loop over the rows of `mystates`, first writing the character value, then the numeric ones:

```
f = file('states.bin','wb')
for(i in 1:nrow(mystates)){
    writeBin(mystates$newname[i],f)
    writeBin(unlist(mystates[i,2:9]),f)
}
```

Note the use of `unlist` to convert a row of the data frame to a vector suitable for `writeBin`.

2.14 Writing R Objects to Files in ASCII Format

While the binary format that R uses to store data (Section 2.12) is the natural choice for saving data that will be used by R, there are several other ways that the contents of R objects can be written to files. The idea of a human-readable (nonbinary) file with data is very attractive, since most programs can read files of this type, and, in the worst possible case, you can see what the file contains by using an ordinary editor. R provides two functions for writing objects to files in ASCII format; `write`, which is suitable for the same kinds of data as `scan` (Section 2.1), and `write.table`, which is suitable for the types of data which would normally be read using `read.table` (Section 2.2).

2.14.1 The `write` Function

The `write` function accepts an R object and the name of a file or connection object, and writes an ASCII representation of the object to the appropriate destination. The `ncolumns=` argument can be used to specify the number of values to write on each line; it defaults to five for numeric variables, and one for character variables. To build up an output file incrementally, the `append=TRUE` argument can be used.

Note that matrices are internally stored by columns, and will be written to any output connection in that order. To write a matrix in row-wise order, use its transpose and adjust the `ncolumn=` argument appropriately. For example, to write the values in the `state.x77` matrix to file in row-wise order, the following statement could be used:

```
> write(t(state.x77),file='state.txt',ncolumns=ncol(state.x77))
```

2.14.2 The `write.table` function

For mixed-mode data, like data frames, the basic tool to produce ASCII files is `write.table`. The only required argument to `write.table` is the name of a dataset or matrix; with just a single argument, the output will be printed on the console, making it easy to test that the file you'll be creating is in the correct format. Usually, the second argument, `file=` will be used to specify the destination as either a character string to represent a file, or a connection (Section 2.1).

By default, character strings are surrounded by quotes by `write.table`; use the `quote=FALSE` argument to suppress this feature. To suppress row names or column names from being written to the file, use the `row.names=FALSE` or `col.names=FALSE` arguments, respectively. Note that `col.names=TRUE` (the default) produces the same sort of headers that are read using the `header=TRUE` argument of `read.table`. Finally, the `sep=` argument can be used to specify a separator other than a blank space. Using `sep=','` (comma-separated) or `sep='\t'` (tab-separated) are two common choices.

For example, to write the `CO2` data frame as a comma-separated file without row names, but with column headers and quotes surrounding character strings, we could use

```
> write.table(CO2,file='co2.txt',row.names=FALSE,sep=',')
```

Similarly to `read.csv` and `read.csv2`, the functions `write.csv` and `write.csv2` are provided as wrappers to `read.table`, with appropriate options set to produce comma- or semicolon-separated files. In addition, the `write.fwf` function in the `gdata` package, available from CRAN, provides a similar functionality for writing R objects to a file using fixed-width fields.

2.15 Reading Data from Other Programs

It sometimes becomes necessary to access data which was created by a program other than R, or to create data in a form that will be easily accessible by some other program. When collaborating with others, they may have already created a saved object using some other program, or may want a dataset you're working with in a format that their favorite program can understand. You may also encounter situations where some other program is more suitable for a particular task, and, once you've created a saved object with that program, you'll want to bring the results into R. In many cases, the most expedient solution is to rely on human-readable comma-separated files to provide access to data, since almost every program can read such files. If this is not an option, or if there are a large number of datasets that need to be processed or created, it may make sense to try to read data directly into R from the file created by the other program, or to write data from R into a format more suitable to some other program.

Function(s)	Purpose
`data.restore`	read `data.dump` output
`read.S`	or saved objects from S version 3
	may work with older Splus objects
`read.dbf`	read or write saved objects
	from DBF files (FoxPro, dBase, etc.)
`read.dta`	read saved objects from Stata (versions 5-9)
`write.dta`	create a Stata saved object
`read.epinfo`	read saved objects from epinfo
`read.spss`	read saved objects from SPSS
	written using the `save` or `export` command
`read.mtp`	read Minitab Portable Worksheet files
`read.octave`	read saved objects from GNU octave
`read.xport`	read saved objects in SAS export format
`read.systat`	read saved objects from systat
	rectangular (mtype=1) data only

Table 2.3. Functions in the `foreign` package

The `foreign` package, available from CRAN, provides programs to read and write data in formats supported by a variety of different programs, summarized in Table 2.3. None of the programs in the table require that the foreign program be available on your computer; for example, you can read and write Stata files (using `read.dta` and `write.dta`) even though you may not have a copy of Stata on your computer. The functions that read files all require a filename argument; those that write files require the data frame to be written as the first argument and a destination filename as the second. Some of the functions listed have additional options to control factor conversion and variable name conventions; full details can be found in their respective help files.

For getting data into other programs, the package also provides the `write.foreign` program, which will generate two files: one containing the data in a form that the foreign program can read, and the second containing instructions that will allow the foreign program to read the data. This provides an alternative means of making data available to someone who wishes to use a program other than R. Currently, `write.foreign` supports SPSS, Stata, and SAS. The help page for `write.foreign` explains how to extend it to support other programs.

To use `write.foreign`, provide the name of a data frame, along with a filename where the data will be written (`datafile=`), and a second filename where the foreign program will be written (`codefile=`), along with the `package=` argument indicating the target program. For example, to create data and programs to read an R data frame called `mydata` into Stata, the following call to `write.foreign` could be used:

```
> write.foreign(mydata,'mydata.txt','mydata.stata',
+                   package='Stata')
```

If mydata.stata is provided as input to Stata, with mydata.txt in the current directory, it will load the data from mydata into Stata.

In the case of SAS, the read.ssd function in the foreign package will create an R data frame from any SAS dataset (not just those in export format), by writing and executing a SAS program to write the data in export format and then calling read.xport. Thus, to use the program, SAS must be available on your computer. If this is the case, the Hmisc package, available from CRAN, also provides a number of programs useful for working with SAS datasets by using SAS to process the data.

3

R and Databases

While many tasks that used to be performed using relational databases can be easily implemented in R, there are some situations where using the power of a relational database nicely complements the capabilities of R. One obvious example are situations where the data to be used is stored in a relational database. Relational databases can also be used to make working with very large datasets easier.

The topic of administration of a database is beyond the scope of this book, and the assumption will always be made that you have access to a running database, and that enough permissions have been granted to perform the necessary database operations.

There are two principal ways to connect with databases in R. The first uses the ODBC (Open DataBase Connectivity) facility available on many computers. The second uses the DBI package of R along with a specialized package for the particular database needed to be accessed. If there is a specialized package available for your database, you may find that the corresponding DBI-based package may give better performance than the ODBC approach. On the other hand, if you are using a database for which a specialized package is not available, using ODBC may be your only option.

3.1 A Brief Guide to SQL

3.1.1 Navigation Commands

Since a single server may hold more than one database, each with potentially many tables, and since each table can contain many columns (variables), it may be useful to examine exactly what's available in a database before starting to work with it. Often there are graphical clients available to communicate with databases that will present this information in a convenient form, but R can also be used to create data frames containing this information. The table below shows some common tasks and the SQL statements to execute them; when used with **dbGetQuery** they will each return a data frame with the requested information. In the table below, keywords are shown in uppercase; the terms in lowercase would be replaced by those specific to your task. When

Task	SQL query
Find names of available databases	`SHOW DATABASES`
Find names of tables in a database	`SHOW TABLES IN database`
Find names of columns in a table	`SHOW COLUMNS IN table`
Find the types of columns in a table	`DESCRIBE table`
Change the default database	`USE database`

Table 3.1. Basic SQL commands

using command-line clients, each SQL statement must end in a semicolon, but the semicolon is not required when using the **RMySQL** interface. Keywords will always be recognized regardless of case, but depending on the version of MySQL that the server is using, database, table, and column names may or may not be case-sensitive.

3.1.2 Basics of SQL

The first step to understanding SQL is to realize that, unlike R, it is not a programming language; operations in SQL are performed using individual queries without loops or control statements. The most important SQL command is SELECT. Since queries are performed using single statements, the syntax of the SELECT command can be quite daunting:

```
SELECT columns or computations
      FROM table
      WHERE condition
      GROUP BY columns
      HAVING condition
      ORDER BY column  [ASC | DESC]
      LIMIT offset,count;
```

Fortunately, most of the clauses in the SELECT statement are optional. In fact, many queries will simply retrieve all of the data in a particular table through the following command:

```
SELECT * FROM tablename;
```

The asterisk (*) means "all the columns in the table". Alternatively, a comma-separated list of variables or expressions can be supplied:

```
SELECT var1,var2,var2/var1 from tablename;
```

will return three columns corresponding to `var1`, `var2`, and the computed value of their ratio. A useful operator in `SQL` is the `AS` operator, which can be used to change the name of a column in the result set. In the previous example, if we wanted the name of the third column to be "`ratio`" we could use the `AS` command:

```
SELECT var1,var2,var2/var1 AS ratio FROM tablename;
```

In fact, the use of the word `AS` is optional; the new column name can simply follow the old one. In these examples, I will include the `AS` keyword, as it makes the query more readable. This same technique can be used to refer to tables through alternative names as well.

To limit the rows which are returned, the `WHERE` clause can be used. Most common operators can be used to define expressions for the `WHERE` clause, along with the keywords `AND` and `OR`. For example, to extract all columns for the rows of a table where `var1` is greater than 10 and `var2` is less than `var1`, we could use

```
SELECT * FROM tablename WHERE var1 > 10 AND var2 < var1;
```

One limitation of the `WHERE` clause is that it cannot access variables that were created in the `SELECT` statement; the `HAVING` clause must be used in those cases. So to find cases where the computed ratio is greater than 10, we could use a statement like this:

```
SELECT var1,var2,var2/var1 AS ratio
    FROM tablename HAVING ratio > 10;
```

Notice the similarity between these simple queries and the `subset` function (Section 6.8).

Two operators in SQL are especially useful for character variables. The `LIKE` operator allows the use of "`%`" to represent zero or more of any character, and "`_`" to represent exactly one character. The `RLIKE` operator allows the use of regular expressions for character comparisons (see Section 7.4).

3.1.3 Aggregation

The `GROUP BY` clause, in conjunction with some SQL-provided aggregation functions, can be useful if you wish to produce a table of counts or a data summary from a database, without bringing all of the data into R. Some of the common aggregation functions available in SQL are summarized in Table 3.2. For example, suppose we wanted to create a table of means for a variable `x`, from a database table named `table`, broken down by a categorical variable called `type`. We could create a table with the value of `type` and `mean` with the following statement:

```
SELECT type,AVG(x) AS mean FROM table GROUP BY type;
```

Task	SQL aggregation function
Count numbers of occurrences	COUNT()
Find the mean	AVG()
Find minimum	MIN()
Find maximum	MAX()
Find variance	VAR_SAMP()
Find standard deviation	STDDEV_SAMP()

Table 3.2. Basic SQL aggregation commands

Remember to include the grouping variable in the list of selected variables, as SQL will not do this automatically. Since `mean` is a calculated variable in this example, you would need to use a `HAVING` clause to limit the observations that were returned based on the value of `mean`.

Since the number of observations for any column in a particular table will always be the same, it is common practice to use an asterisk (∗) as an argument to the `COUNT` aggregation function. To create a table of counts by `type` in the previous example, we could use

```
SELECT type,COUNT(*) FROM table GROUP BY type;
```

To group by more than one variable, use a comma-separated list as an argument to the `GROUP BY` clause.

Multiple aggregated statistics can easily be output in a single query. Suppose we wanted to count the number of observations for each `type`, along with the mean and standard deviation of the x column. The following command could be used:

```
SELECT type,COUNT(*),AVG(x) AS mean,STDDEV_SAMP(x) AS std
    FROM accounts GROUP BY type;
```

3.1.4 Joining Two Databases

One of the strengths of database servers is that they can effectively join together multiple database tables, based on common values of columns within the tables. Of course, the same capability is available within R through the `merge` function (see Section 9.6), but it may be more efficient to use the database server for merging.

The most common way of joining two tables is through an inner join; only those observations that have common values of the variable used for the merge will be retained in the output table. (This is also the default behavior of the `merge` function.) For example, suppose we have a table called `children` with columns `id`, `family_id`, and `height` and `weight`, and a second table called `mothers`, with columns `id`, `family_id`, and `income`. We would like a table with the `height` and `weight` of the children, along with the income of the mothers. The following SQL statement will return the table:

```
SELECT height,weight,income FROM children
   INNER JOIN mothers USING(family_id);
```

The variable in the USING expression (family_id in this example) is known as a key or sometimes a foreign key. If the two tables being joined have only one variable in common, the INNER JOIN can be replaced with a NATURAL JOIN, and the USING expression can be omitted.

Now suppose we wish to produce a table with both the children's id and the mother's id. Since there are variables called id in both data tables, we need to distinguish between them by preceding the column name with the table name and a period. In this example, we could use a query like this:

```
SELECT children.id,mothers.id,height,weight,income
   FROM children INNER JOIN mothers USING(family_id);
```

The AS operator can be used to make it easier to refer to multiple tables, as well as renaming the columns:

```
SELECT c.id as kidid,m.id as momid,height,weight,income
   FROM children AS c
   INNER JOIN mothers AS m USING(family_id);
```

3.1.5 Subqueries

Continuing with the current example, consider the task of tabulating the family size (i.e. the number of children with the same family_id) for all the families in the database. It's easy to create a table that has the counts for each value of family_id:

```
SELECT family_id,COUNT(*) AS ct FROM children
   GROUP BY family_id;
```

How can we then count how many of each family size was found? One way would be to create a temporary table containing the ids and sizes, and querying that table, but often the permission to create new tables on a server is not available. The alternative is to use subqueries. In SQL, a subquery is a query surrounded by parentheses, which can be treated just like any other table. One restriction of subqueries is that all subquery tables must be given an alias (through the AS operator), even if you won't be directly referring to the table. We can produce the table of family sizes with the following query:

```
SELECT ct,COUNT(*) as n
   FROM (SELECT COUNT(*) AS ct FROM children
         GROUP BY family_id) AS x
   GROUP BY ct;
```

Subqueries are also useful when the timing of database operations makes a query impossible for the database to understand. Let's say we wanted all the available information about the tallest child in the database. One obvious possibility is to perform the following query:

```
SELECT * FROM children WHERE height = MAX(height);
```

Depending on the database you use, you might get an empty set, or a syntax error. To get around the problem, we can create a table with only the maximum height, and then use it in a subquery:

```
SELECT * FROM children
    WHERE height = (SELECT MAX(height) as height from children);
```

3.1.6 Modifying Database Records

To change the values of selected records in a database, the UPDATE command can be used. The format of the UPDATE statement is

```
UPDATE table SET var=value
    WHERE condition
    LIMIT n;
```

To change more than one variable's value, the var=value specification can be replaced with a comma-separated list of variable/value pairs. The WHERE and LIMIT specifications are optional. If a LIMIT specification is provided, only that many records will be considered for updating, even if some of the chosen records will not actually be modified. For example, to change the height and weight for a subject with a particular id, we could use a statement like

```
UPDATE children SET weight=100,height=55
    WHERE id = 12345;
```

To completely remove a record, the DELETE statement can be used. The basic syntax is as follows:

```
DELETE FROM table
    WHERE condition
    LIMIT n;
```

Without a WHERE clause, all of the records of the database table will be removed, so this statement should be used with caution. If a LIMIT specification is provided, it will be based on observations matching the condition of the WHERE clause, if one is specified.

Finally, to completely remove an entire table or database, the DROP statement can be used, for example

```
DROP TABLE tablename;
```

or

```
DROP DATABASE dbname;
```

When using the DROP command, an error will be reported if the table or database to be dropped does not exist. To avoid this, the IF EXISTS clause can be added to the DROP statement, as in

```
DROP DATABASE IF EXISTS dbname;
```

Notice that these commands take effect on the database as soon as they are issued, so it's a good idea to have a backup of the data in the database before using these commands.

3.2 ODBC

The ODBC (Open DataBase Connectivity) facility allows access to a variety of databases through a common interface. In R, the RODBC package, available from CRAN, is used to access this capability. ODBC was originally developed on Windows, and the widest variety of ODBC connectors will be available on that platform. However, both Linux and Mac OS X also provide database connectivity through ODBC. If you need to use a database in R that is not directly supported, RODBC will probably be your best choice, as many database manufacturers provide ODBC connectors for their products.

The first step in using RODBC is to set up a DSN or data source name. In order to do this, you need to know the name that your computer uses for a particular data source. On Windows, the ODBC Source Administrator (accessed through Control Panel → Administrative Tools → Data Sources(ODBC)) is used to establish DSNs. Under the "Drivers" tab, you can see what connectors are available on your computer, and the name that is used to access them. If you install additional connectors, you should see them listed here. You can use this name, providing additional connection details each time you create a connection, or you can create a new DSN to automate the process. To create a new DSN, click on the "Add" button under the User DSN tab, choose an appropriate driver from the pop-up window, and click "Finish". At this point a dialog specific to the database you're using will appear, and you can fill in the required information to create the DSN. Make sure to note the name that you use for the data source name, since that is how the ODBC connection is specified.

Under Mac OS X, the ODBC Administrator (which can be found in the /Applications/Utilities folder) performs a similar function. You can view available drivers in the "Drivers" tab, or choose the User DSN tab, and click "Add" to create a new DSN; after choosing a driver, you can configure it using keyword/value pairs appropriate for the particular database you are using.

To use the RODBC package on a Linux system, the unixodbc libraries must first be installed. Most linux distributions will make this very easy. The configuration of UNIXODBC is controlled by two files: odbcinst.ini and odbc.ini. The first file contains the available ODBC drivers, and the second file is used to define additional DSNs, if desired. For example, the following is an odbc.ini file which defines a DSN called myodbc using the MySQL ODBC driver:

```
[myodbc]
Driver       = MySQL
Description  = MySQL ODBC 2.50 Driver DSN
Server       = localhost
Port         = 3306
User         = user
Password     = password
Database     = test
```

The name in square brackets (myodbc in this case) is the DSN that is being defined; multiple DSNs can be defined in a single file by starting a new section with the DSN in brackets. In order to use a driver, it must be defined in the odbcinst.ini file. The specific keywords in the file will depend on the specific connector being used.

By default on most systems, the two configuration files for UNIXODBC are in the /etc directory. To specify a different location for odbc.ini, set the environmental variable ODBCINI to the fully-qualified filename of the file; to specify a different location for odbcinst.ini, set the environmental variable ODBCSYSINI to the **directory** in which odbcinst.ini can be found.

3.3 Using the RODBC Package

After loading the RODBC package, if you've configured a DSN that provides all the necessary information to connect and access your database, you can create a connection by simply passing the DSN to the odbcConnect function. Suppose we have a DSN named myodbc to connect to a MySQL database, and we have provided the server, username, password, and database in the DSN definition. Then we can create a connection through RODBC as follows:

```
> library(RODBC)
> con = odbcConnect('myodbc')
```

Additional keywords defining the connection can be provided in the DSN argument by separating keyword=value pairs with semicolons. For example, if a DSN was created without specifying a required password, the database could be accessed as follows:

```
> con = odbcConnect('myodbc;password=xxxxx')
```

Other possible keywords depend on the particular data source. For MySQL these keywords include server, user, password, port, and database; for PostgreSQL, substitute username for user.

Once you've got a connection to the ODBC source, the sqlQuery function allows any valid SQL query to be sent to the connection. This will be the case even if SQL is not the native language of the underlying database. Passed only a connection and a query, sqlQuery will return a data frame containing the entire result of the query. The `max=` argument to `sqlQuery` will limit the number of rows returned, and can be followed by repeated calls to `sqlGetResults` (also using appropriate `max=` arguments) to process a query in smaller pieces.

To prevent unnecessary resource use, the `odbcClose` function should be passed any ODBC connection objects when they are no longer needed.

3.4 The DBI Package

One of the most popular databases used with R is MySQL (`http://mysql.com`). This freely available database runs on a variety of platforms and is relatively easy to configure and operate.

In the following sections, we'll look at the RMySQL package as an example of using the DBI package.

3.5 Accessing a MySQL Database

The first step in accessing a MySQL database is loading the MySQL package. This package will automatically load the required DBI package, which provides a common interface across different databases. Next, the MySQL driver is loaded via the `dbDriver` function, so that the DBI interface will know what type of database it's communicating with:

```
> library(RMySQL)
> drv = dbDriver("MySQL")
```

Now, the specifics of the database connection can be provided through the `dbConnect` function. These include the database name, the database username and password, and the host on which the database is running. If the database is running on the same machine as your R session, the hostname can be omitted. For example, to access a database called "`test`", via a user name of "sqluser" and password of "secret" on the host "`sql.company.com`", the following call to `dbConnect` could be used:

```
> con = dbConnect(drv,dbname='test',user='sqluser',
+                  password='secret',host='sql.company.com')
```

The calls to `dbDriver` and `dbConnect` need only be made once for an entire session. Note that the `dbname` passed to `dbConnect` might represent a collection of many tables; the specific table to be used will be specified in the queries that are sent to the database.

You can close an unused DBI connection by passing the connection object to `dbDisconnect`.

3.6 Performing Queries

SQL queries make requests for some or all of the variables in one or more database tables, so a natural way to package these results within R is in a data frame. In most cases, a single call to dbGetQuery can be used to send a query to the database, and have the resulting table returned as a data frame. For example, suppose that we have connected to the database "test" as described in the previous section, and we wish to extract all of the observations in a table called "mydata". After the appropriate calls to dbDriver and dbConnect, we could retrieve the data with the following command:

```
> mydata = dbGetQuery(con,'select * from mydata')
```

Any valid SQL query can be passed to a database by this method.

In the case where data needs to be processed in pieces, the dbSendQuery function can be used to initiate the query, and the fetch function can be passed the result from dbSendQuery to sequentially access the results of the query. Once all the required data is extracted using fetch, the result from dbSendQuery should be passed to the dbClearResult function to insure that the next query will be properly processed. (When using dbGetQuery there is no need to call an additional function at the end of the query.) Note that by default, the fetch function will return 500 records at a time; this can be overwritten with the n= argument, using a value of -1 to indicate all of the available records, or an integer to specify the number of records desired.

3.7 Normalized Tables

The principle of normalization is central to database design. The goal of normalization is to eliminate redundancy in the information stored in the database tables. To achieve this goal, what might be a single data frame in R might be broken into several tables in a database. For example, suppose we are working with a database containing information about the parts required to produce a product. If we stored the part name, supplier's name, and the price of the part all in one database, we would have many records with identical information about suppliers. In a properly normalized database, there would be two tables; one with part names and prices and an id representing the supplier of the part. This id, known as a key or foreign key, would also be found exactly once in a second table containing supplier information. Suppose the first table is called parts with columns name, price, and supplierid, and the second table is called suppliers, with columns supplierid and name. Our goal is to create a data frame with the part name and price along with the name of the supplier. An appropriate query to retrieve the table we want into a data frame could be written as

```
> result = dbGetQuery(con,'SELECT parts.name,parts.description,
+                            supplier.name AS supplier
+                            FROM parts INNER JOIN
+                            suppliers USING(supplierid)')
```

Using the database to merge the tables makes sense if you're familiar with SQL, and especially if the tables you're working with are very large. However, the tables could also be retrieved in their entirety, and the merging performed in R:

```
> parts = dbGetQuery(con,'SELECT * FROM parts')
> suppliers = dbGetQuery(con,'SELECT * FROM suppliers')
> result = merge(parts,suppliers,by='supplierid')
```

This simplistic solution, while workable, ignores the motivation behind the initial normalization of the database tables, namely, to avoid redundancy. The supplier name variable is being stored as a character variable, whose value is repeated in the result data frame for each observation from the same supplier. A more efficient solution is to note the similarity between the suppliers table and the idea of a factor in R. The supplierids represent the levels of a factor, and the names represent the labels. Thus, we can create a data frame storing the suppliers as a factor with code like this:

```
> parts = dbGetQuery(con,'SELECT * FROM parts')
> suppliers = dbGetQuery(con,'SELECT * FROM suppliers')
> result = data.frame(name=parts$name,price=parts$price,
+                        supplier=factor(parts$supplierid,
+                        levels=suppliers$supplierid,
+                        labels=suppliers$name))
```

Since the data.frame function automatically converts character variables to factors, both name and supplier will be stored as factors.

3.8 Getting Data into MySQL

If your data is already in an R object, it can be easily transfered to a database using the dbWriteTable function, which accepts the same sort of connection object that dbGetQuery uses. By using the append=TRUE argument to dbWriteTable, a large database table can be built using smaller pieces.

   If it is desired to create a table directly from raw data, it is first necessary to describe the nature of each column in the table with the CREATE TABLE statement. For example, one way of creating a table called mydata to hold columns name (a character variable) and number (a floating point value) would be to issue a statement like:

```
CREATE TABLE mydata (name text, number double);
```

This statement could be submitted to MySQL by, for example, passing it to dbGetQuery (although it will not return any value). To make generating statements like this easier, the dbBuildTableDefinition function can be used; it will generate appropriate statements to create a database suitable to hold an R data frame. Following the current example, we could generate the CREATE TABLE statements in R as follows:

```
> x = data.frame(name='',number=0.)
> cat(dbBuildTableDefinition(dbDriver('MySQL'),
+                             'mydata',x),"\n")
CREATE TABLE mydata
( row_names text,
        name text,
        number double
)
```

To suppress the row_names column, the row.names=FALSE argument can be used. The output from dbBuildTableDefinition can be passed directly to dbGetQuery to create the table in the database. If you wish to create a table with identical specifications to an existing table, the LIKE clause can be used in the CREATE TABLE statement, as in

```
CREATE TABLE newtable LIKE oldtable;
```

To get an understanding of how existing tables are stored in the database, the DESCRIBE table statement can be used.

Once the table has been created, the actual data needs to be entered. The SQL INSERT command can be used to add one or more observations to a database table. When the columns defined by the CREATE TABLE command are being entered in the order they are stored in the database table, all that is required is the VALUES keyword:

```
INSERT INTO mydata VALUES('fred',7);
```

If the values are to be entered in an order different from how they are stored in the database table, a parenthetical comma-separated list describing the order that will be used needs to be provided before the VALUES keyword. So to add an observation only specifying the number value before the name value, we could use the following SQL command:

```
INSERT INTO mydata (number,name) values(7,'fred');
```

To add additional observations, additional parenthesized comma-separated lists, themselves separated by commas, can be added at the end of the INSERT command. The following command adds two new observations to the mydata table:

```
INSERT INTO mydata VALUES('tim',12),('sue',9);
```

Generally, however, it will be advantageous to insert all of the data into the database in a single database call, either through an external program or through the LOAD DATA command. With MySQL, the mysqlimport shell command can be used to read whole files of data into a database table. Among its arguments are --local, which specifies that the data is local, and not on the server, --delete, which insures that the contents of any current table with the same name are removed before creating the new table, and --fields-terminated-by= and --lines-terminated-by= to provide the field and line terminators, respectively. In addition to these optional arguments, the -u username option, to provide the MySQL username, the -h hostname option to provide the name of the machine on which the MySQL server is running, and the -p option, to tell the server to prompt for a password, may be required to establish a database connection. In addition, since the MySQL server won't read header lines, the --ignore-lines=1 argument can be used to skip a header line.

For example, to read a comma-separated text file called mydata.txt into a mysql database called test, the following shell command could be entered in a terminal window:

```
mysqlimport -u sqluser -p --delete --local \
            --fields-terminated-by=',' test mydata.txt
```

Notice that mysqlimport determines the table name by removing any suffix from the name of the file containing the data (mydata.txt in this example). As with the LOAD DATA command, the table to hold the data must be created before mysqlimport can be used.

The same operation can be performed by sending MySQL statements to the server. Assuming an appropriate connection object has been obtained, we could load data from the mydata.txt file into the database with the following call to dbGetQuery:

```
> dbGetQuery(con,"LOAD DATA INFILE 'mydata.txt'\
+            INTO TABLE mydata\  FIELDS TERMINATED BY ','")
```

Once the data is loaded into the database, the SELECT statement can be used to create subsets of the data which will be manageable inside of R.

3.9 More Complex Aggregations

The dbApply function can be used to apply a user-specified R function to groups of data extracted from a database. To use dbApply, first create a result set object through a call to dbSendQuery, using the ORDER BY clause to insure that the data will be brought into R in the appropriate order. The result set object can then be passed to dbApply, along with an INDEX= argument to specify the grouping variable, and a FUN= argument, to specify the function to be applied to each group. This function must accept two arguments: the first is the data frame consisting of the requested data for a given group, and the second is the value of the grouping variable. For example, suppose we have a database table called `cordata` with columns `group`, `x`, and `y`, and we wish to find the correlation between `x` and `y`, broken down by groups. First, we use `dbSendQuery` to create the result set object:

```
> res1 = dbSendQuery(con,
+                    'SELECT group,x,y FROM cordata ORDER BY group')
```

Now we can pass this result set object to `dbApply` to obtain the result:

```
> correlations = dbApply(res1,INDEX='group',
+                        FUN=function(df,group)cor(df$x,df$y))
```

The return value from `dbApply`, `correlations`, will be a list of correlations whose names represent the levels of the `group` variable.

If the `dbApply` function is not available for a particular database, or if more control is required over the aggregation, the following function shows an alternative means of applying a function to subsets of the data:

```
mydbapply = function(con,table,groupv,otherv,fun){
    query = paste('select ',groupv,' from ',table,
                  ' group by ',groupv,sep='')
    queryresult = dbGetQuery(con,query)
    answer = list()
    k = 1
    varlist = paste(c(groupv,otherv),collapse=',')
    for(gg in queryresult[[groupv]]){
        qry = paste('select ',varlist,' from ',table,'
                    where ', groupv,' = "',gg,'"',sep='')
        qryresult = dbGetQuery(con,qry)
        answer[[k]] = fun(qryresult)
        names(answer)[k] = as.character(gg)
        k = k + 1
    }
    return(answer)
}
```

The arguments to `mydbapply` are `con`, an active database connection object, `groupv`, a character string representing the database column to be used for grouping, `otherv`, a character vector containing the names of other database columns that need to be extracted from the database, and `fun`, the function that will operate on the data frame containing the grouping and other variables. The example of the previous section could be executed using `mydbapply` as

```
> correlations = mydbapply(con,'cordata','group',c('x','y'),
                  function(df)cor(df$x,df$y))
```

4

Dates

R provides several options for dealing with date and date/time data. The built-in `as.Date` function handles dates (without times); the contributed package `chron` handles dates and times, but does not control for time zones; and the `POSIXct` and `POSIXlt` classes allow for dates and times with control for time zones. The general rule for date/time data in R is to use the simplest technique possible. Thus, for date only data, `as.Date` will usually be the best choice. If you need to handle dates and times, without time-zone information, the `chron` package is a good choice; the POSIX classes are especially useful when time-zone manipulation is important. Also, don't overlook the various "as." functions (like `as.Date` and `as.POSIXlt`) for converting among the different date types when necessary.

Except for the `POSIXlt` class, dates are stored internally as the number of days or seconds from some reference date. Thus, dates in R will generally have a numeric mode, and the `class` function can be used to find the way they are actually being stored. The `POSIXlt` class stores date/time values as a list of components (`hour`, `min`, `sec`, `mon`, etc.) making it easy to extract these parts.

To get the current date, the `Sys.Date` function will return a `Date` object which can be converted to a different class if necessary.

The following sections will describe the different types of date values in more detail.

4.1 as.Date

The `as.Date` function allows a variety of input formats through the `format=` argument. The default format is a four-digit year, followed by a month, then a day, separated by either dashes or slashes. Some examples of dates which `as.Date` will accept by default are as follows:

```
> as.Date('1915-6-16')
[1] "1915-06-16"
```

```
> as.Date('1990/02/17')
[1] "1990-02-17"
```

Code	Value
%d	Day of the month (decimal number)
%m	Month (decimal number)
%b	Month (abbreviated)
%B	Month (full name)
%y	Year (2 digit)
%Y	Year (4 digit)

Table 4.1. Format codes for dates

If your input dates are not in the standard format, a format string can be composed using the elements shown in Table 4.1. The following examples show some ways that this can be used:

```
> as.Date('1/15/2001',format='%m/%d/%Y')
[1] "2001-01-15"
> as.Date('April 26, 2001',format='%B %d, %Y')
[1] "2001-04-26"
> as.Date('22JUN01',format='%d%b%y')
[1] "2001-06-22"
```

Internally, Date objects are stored as the number of days since January 1, 1970, using negative numbers for earlier dates. The as.numeric function can be used to convert a Date object to its internal form. To convert this form back to a Date object, it can be assigned a class of Date directly:

```
> thedate = as.Date('1/15/2001',format='%m/%d/%Y')
> ndate = as.numeric(thedate)
> ndate
[1] 11337
> class(ndate) = 'Date'
> ndate
[1] "2001-01-15"
```

To extract the components of the dates, the weekdays, months, days, or quarters functions can be used. For example, to see if the R developers favor a particular day of the week for their releases, we can first extract the release dates from the CRAN website with a program like this:

```
f = url('http://cran.cnr.berkeley.edu/src/base/R-2','r')
rdates = data.frame()
while(1){
    l = readLines(f,1)
    if(length(l) == 0)break
    if(regexpr('href="R-',l) > -1){
            parts = strsplit(l,' ')[[1]]
            rver = sub('^.*>(R-.*).tar.gz.*','\\1',l)
        date = parts[18]
        rdates = rbind(rdates,data.frame(ver=rver,Date=date))
        }
}
rdates$Date = as.Date(rdates$Date,'%d-%B-%Y')
```

Then, the days of the week can be tabulated after using the weekdays function as follows:

```
> table(weekdays(rdates$Date))

 Monday Thursday  Tuesday
      5        3        4
```

Monday, Thursday, and Tuesday seem to be the favorite days for releases.

For an alternative way of extracting pieces of a date, and for information on possible output formats for Date objects, see Section 4.3.

4.2 The chron Package

The chron function converts dates and times to chron objects. The dates and times are provided to the chron function as separate values, so some preprocessing may be necessary to prepare input date/times for the chron function. When using character values, the default format for dates is the decimal month value followed by the decimal day value followed by the year, using the slash as a separator. Alternative formats can be provided by using the codes shown in Table 4.2.

Alternatively, dates can be specified by a numeric value, representing the number of days since January 1, 1970. To input dates stored as the day of the year, the origin= argument can be used to interpret numeric dates relative to a different date.

The default format for times consists of the hour, minutes, and seconds, separated by colons. Alternative formats can use the codes in Table 4.2.

Often the first task when using the chron package is to break apart the date and times if they are stored together. In the following example, the strsplit function is used to break apart the string.

Format codes for dates	
Code	Value
m	Month (decimal number)
d	Day of the month (decimal number)
y	Year (4 digit)
mon	Month (abbreviated)
month	Month (full name)
Format codes for times	
Code	Value
h	Hour
m	Minute
s	Second

Table 4.2. Format codes for chron objects

```
> library(chron)
> dtimes = c("2002-06-09 12:45:40","2003-01-29 09:30:40",
+            "2002-09-04 16:45:40","2002-11-13 20:00:40",
+            "2002-07-07 17:30:40")
> dtparts = t(as.data.frame(strsplit(dtimes,' ')))
> row.names(dtparts) = NULL
> thetimes = chron(dates=dtparts[,1],times=dtparts[,2],
+                  format=c('y-m-d','h:m:s'))
> thetimes
[1] (02-06-09 12:45:40) (03-01-29 09:30:40) (02-09-04 16:45:40)
[4] (02-11-13 20:00:40) (02-07-07 17:30:40)
```

Chron values are stored internally as the fractional number of days from January 1, 1970. The as.numeric function can be used to access the internal values.

If times are stored as the number of seconds since midnight, they can be accommodated by the POSIX classes (see Section 4.3).

For information on formatting chron objects for output, see Section 4.3.

4.3 POSIX Classes

POSIX represents a portable operating system interface, primarily for UNIX systems, but available on other operating systems as well. Dates stored in the POSIX format are date/time values (like dates with the chron package), but also allow modification of time zones. Unlike the chron package, which stores times as fractions of days, the POSIX date classes store times to the nearest second, so they provide a more accurate representation of times.

There are two POSIX date/time classes, which differ in the way that the values are stored internally. The POSIXct class stores date/time values as the number of seconds since January 1, 1970, while the POSIXlt class stores them as a list with elements for second, minute, hour, day, month, and year, among others. Unless you need the list nature of the POSIXlt class, the POSIXct class is the usual choice for storing dates in R.

The default input format for POSIX dates consists of the year, followed by the month and day, separated by slashes or dashes; for date/time values, the date may be followed by white space and a time in the form hour:minutes:seconds or hour:minutes; thus, the following are examples of valid POSIX date or date/time inputs:

```
1915/6/16
2005-06-24 11:25
1990/2/17 12:20:05
```

If the input times correspond to one of these formats, as.POSIXct can be called directly:

```
> dts = c("2005-10-21 18:47:22","2005-12-24 16:39:58",
+         "2005-10-28 07:30:05 PDT")
> as.POSIXlt(dts)
[1] "2005-10-21 18:47:22" "2005-12-24 16:39:58"
[3] "2005-10-28 07:30:05"
```

If your input date/times are stored as the number of seconds from January 1, 1970, you can create POSIX date values by assigning the appropriate class directly to those values. Since many date manipulation functions refer to the POSIXt pseudo-class, be sure to include it in the class attribute of the values.

```
> dts = c(1127056501,1104295502,1129233601,1113547501,
+         1119826801,1132519502,1125298801,1113289201)
> mydates = dts
> class(mydates) = c('POSIXt','POSIXct')
> mydates
[1] "2005-09-18 08:15:01 PDT" "2004-12-28 20:45:02 PST"
[3] "2005-10-13 13:00:01 PDT" "2005-04-14 23:45:01 PDT"
[5] "2005-06-26 16:00:01 PDT" "2005-11-20 12:45:02 PST"
[7] "2005-08-29 00:00:01 PDT" "2005-04-12 00:00:01 PDT"
```

Conversions like this can be done more succinctly using the structure function:

```
> mydates = structure(dts,class=c('POSIXt','POSIXct'))
```

The POSIX date/time classes take advantage of the POSIX date/time implementation of your operating system, allowing dates and times in R to be manipulated in the same way they would be in, for example, a C program. The two most important functions in this regard are strptime, for inputting dates, and strftime, for formatting dates for output. Both of these functions use a variety of formatting codes, some of which are listed in Table 4.3, to

Code	Meaning	Code	Meaning
%a	Abbreviated weekday	%A	Full weekday
%b	Abbreviated month	%B	Full month
%c	Locale-specific date and time	%d	Decimal date
%H	Decimal hours (24 hour)	%I	Decimal hours (12 hour)
%j	Decimal day of the year	%m	Decimal month
%M	Decimal minute	%p	Locale-specific AM/PM
%S	Decimal second	%U	Decimal week of the year (starting on Sunday)
%w	Decimal weekday (0=Sunday)	%W	Decimal week of the year (starting on Monday)
%x	Locale-specific date	%X	Locale-specific time
%y	2-digit year	%Y	4-digit year
%z	Offset from GMT	%Z	Time zone (character)

Table 4.3. Format codes for `strftime` and `strptime`

specify the way dates are read or printed. For example, dates in many logfiles are printed in a format like "16/Oct/2005:07:51:00". To create a `POSIXct` date from a date in this format, the following call to `strptime` could be used:

```
> mydate = strptime('16/Oct/2005:07:51:00',
+                    format='%d/%b/%Y:%H:%M:%S')
[1] "2005-10-16 07:51:00"
```

Note that nonformat characters (like the slashes) are interpreted literally.

When using `strptime`, an optional time zone can be specified with the `tz=` option.

Since POSIX date/time values are stored internally as the number of seconds since January 1, 1970, they can easily use times that are not represented by a formatted version of the hour, minute, and second. For example, suppose we have a vector of date/time values stored as a date followed by the number of seconds since midnight:

```
> mydates = c('20060515 112504.5','20060518 101000.3',
+             '20060520 20035.1')
```

The first step is to split the dates and times, and then use `strptime` to convert the date to a POSIXct value. Then, the times can simply be added to this value:

```
> dtparts = t(as.data.frame(strsplit(mydates,' ')))
> dtimes = strptime(dtparts[,1],format='%Y%m%d') +
+                    as.numeric(dtparts[,2])
> dtimes
[1] "2006-05-16 07:15:04 PDT" "2006-05-19 04:03:20 PDT"
[3] "2006-05-20 05:33:55 PDT"
```

Another way to create POSIX dates is to pass the individual components of the time to the `ISOdate` function. Thus, the first date/time value in the previous example could also be created with a call to `ISOdate`:

```
> ISOdate(2006,5,16,7,15,04,tz="PDT")
[1] "2006-05-16 07:15:04 PDT"
```

ISOdate will accept both numeric and character arguments.

For formatting dates for output, the format function will recognize the type of your input date, and perform any necessary conversions before calling strftime, so strftime rarely needs to be called directly. For example, to print a date/time value in an extended format, we could use:

```
> thedate = ISOdate(2005,10,21,18,47,22,tz="PDT")
> format(thedate,'%A, %B %d, %Y %H:%M:%S')
[1] "Friday, October 21, 2005 18:47:22"
```

When using POSIX dates, the optional usetz=TRUE argument to the format function can be specified to indicate that the time zone should be displayed. Additionally, as.POSIXlt and as.POSIXct can also accept Date or chron objects, so they can be input as described in the previous sections and converted as needed. Conversion between the two POSIX forms is also possible.

The individual components of a POSIX date/time object can be extracted by first converting to POSIXlt if necessary, and then accessing the components directly:

```
> mydate = as.POSIXlt('2005-4-19 7:01:00')
> names(mydate)
[1] "sec"   "min"   "hour"  "mday"  "mon"   "year"
[7] "wday"  "yday"  "isdst"
> mydate$mday
[1] 19
```

4.4 Working with Dates

Many of the statistical summary functions, like mean, min, max, etc are able to transparently handle date objects. For example, consider the release dates of various versions of R from 1.0 to 2.0:

```
> rdates = scan(what="")
1: 1.0 29Feb2000
3: 1.1 15Jun2000
5: 1.2 15Dec2000
7: 1.3 22Jun2001
9: 1.4 19Dec2001
11: 1.5 29Apr2002
13: 1.6 1Oct2002
15: 1.7 16Apr2003
17: 1.8 8Oct2003
19: 1.9 12Apr2004
21: 2.0 4Oct2004
23:
```

```
Read 22 items
> rdates = as.data.frame(matrix(rdates,ncol=2,byrow=TRUE))
> rdates[,2] = as.Date(rdates[,2],format='%d%b%Y')
> names(rdates) = c("Release","Date")
> rdates
     Release      Date
1        1.0 2000-02-29
2        1.1 2000-06-15
3        1.2 2000-12-15
4        1.3 2001-06-22
5        1.4 2001-12-19
6        1.5 2002-04-29
7        1.6 2002-10-01
8        1.7 2003-04-16
9        1.8 2003-10-08
10       1.9 2004-04-12
11       2.0 2004-10-04
```

Once the dates are properly read into R, a variety of calculations can be performed:

```
> mean(rdates$Date)
[1] "2002-05-19"
> range(rdates$Date)
[1] "2000-02-29" "2004-10-04"
> rdates$Date[11] - rdates$Date[1]
Time difference of 1679 days
```

4.5 Time Intervals

If two times (using any of the date or date/time classes) are subtracted, R will return the result in the form of a time difference, which represents a difftime object. For example, New York City experienced a major blackout on July 13, 1977, and another on August 14, 2003. To calculate the time interval between the two blackouts, we can simply subtract the two dates, using any of the classes that have been introduced:

```
> b1 = ISOdate(1977,7,13)
> b2 = ISOdate(2003,8,14)
> b2 - b1
Time difference of 9528 days
```

If an alternative unit of time was desired, the difftime function could be called, using the optional units= argument with any of the following values: "auto", "secs", "mins", "hours", "days", or "weeks". So to see the difference between blackouts in terms of weeks, we can use

```
> difftime(b2,b1,units='weeks')
Time difference of 1361.143 weeks
```

Although `difftime` values are displayed with their units, they can be manipulated like ordinary numeric variables; arithmetic performed with these values will retain the original units.

To convert a time difference in days to one of years, a good approximation is to divide the number of days by 365.25. However, the `difftime` value will display the time units as days. To modify this, the `units` attribute of the object can be modified:

```
> ydiff = (b2 - b1) / 365.25
> ydiff
Time difference of 26.08624 days
> attr(ydiff,'units') = 'years'
> ydiff
Time difference of 26.08624 years
```

4.6 Time Sequences

The `by=` argument to the `seq` function can be specified either as a `difftime` value, or in any units of time that the `difftime` function accepts, making it very easy to generate sequences of dates. For example, to generate a vector of ten dates, starting on July 4, 1976, with an interval of one day between them, we could use

```
> seq(as.Date('1976-7-4'),by='days',length=10)
 [1] "1976-07-04" "1976-07-05" "1976-07-06"
 [4] "1976-07-07" "1976-07-08" "1976-07-09"
 [7] "1976-07-10" "1976-07-11" "1976-07-12"
[10] "1976-07-13"
```

All the date classes except for `chron` will accept an integer before the interval provided as a `by=` argument. We could create a sequence of dates separated by two weeks from June 1, 2000, to August 1, 2000, as follows:

```
> seq(as.Date('2000-6-1'),to=as.Date('2000-8-1'),by='2 weeks')
[1] "2000-06-01" "2000-06-15" "2000-06-29" "2000-07-13"
[5] "2000-07-27"
```

The `cut` function also understands units of `days`, `weeks`, `months`, and `years`, making it very easy to create factors grouped by these units. See Section 5.5 for details.

Format codes can also be used to extract parts of dates, as an alternative to the `weekdays` and other functions described in Section 4.3. We could look at the distribution of weekdays for the R release dates as follows:

```
> table(format(rdates$Date,'%A'))
```

```
Monday Thursday  Tuesday
     5        3        4
```

This same technique can be used to convert dates to factors. For example, to create a factor based on the release dates broken down by years we could use

```
> fdate = factor(format(rdates$Date,'%Y'))
> fdate
 [1] 2004 2004 2005 2005 2005 2005 2006 2006 2006 2006
     2007 2007
Levels: 2004 2005 2006 2007
```

5

Factors

Conceptually, factors are variables in R which take on a limited number of different values; such variables are often referred to as categorical variables. One of the most important uses of factors is in statistical modeling; since categorical variables enter into statistical models differently than continuous variables, storing data as factors insures that the modeling functions will treat such data correctly.

5.1 Using Factors

Factors in R are stored as a vector of integer values with a corresponding set of character values to use when the factor is displayed. The `factor` function is used to create a factor. The only required argument to `factor` is a vector of values which will be returned as a vector of factor values. Both numeric and character variables can be made into factors, but a factor's levels will always be character values. You can see the possible levels for a factor by calling the `levels` function; the `nlevels` function will return the number of levels of a factor.

To change the order in which the levels will be displayed from their default sorted order, the `levels=` argument can be given a vector of all the possible values of the variable in the order you desire. If the ordering should also be used when performing comparisons, use the optional `ordered=TRUE` argument. In this case, the factor is known as an ordered factor.

The levels of a factor are used when displaying the factor's values. You can change these levels at the time you create a factor by passing a vector with the new values through the `labels=` argument. Note that this actually changes the internal levels of the factor, and to change the labels of a factor after it has been created, the assignment form of the `levels` function is used. To illustrate this point, consider a factor taking on integer values which we want to display as roman numerals:

```
> data = c(1,2,2,3,1,2,3,3,1,2,3,3,1)
> fdata = factor(data)
> fdata
 [1] 1 2 2 3 1 2 3 3 1 2 3 3 1
Levels: 1 2 3
> rdata = factor(data,labels=c("I","II","III"))
> rdata
 [1] I   II  II  III I   II  III III I   II  III III I
Levels: I II III
```

To convert the default factor fdata to roman numerals, we use the assignment form of the levels function:

```
> levels(fdata) = c('I','II','III')
> fdata
 [1] I   II  II  III I   II  III III I   II  III III I
Levels: I II III
```

Factors represent a very efficient way to store character values, because each unique character value is stored only once, and the data itself is stored as a vector of integers. Because of this, read.table will automatically convert character variables to factors unless the as.is=TRUE or stringsAsFactors=FALSE arguments are specified, or the stringsAsFactors system option is set to FALSE. See Section 2.2 for details.

As an example of an ordered factor, consider data consisting of the names of months:

```
> mons = c("March","April","January","November","January",
+ "September","October","September","November","August",
+ "January","November","November","February","May","August",
+ "July","December","August","August","September","November",
+ "February","April")
> mons = factor(mons)
> table(mons)
mons
    April   August December February  January     July
        2        4        1        2        3        1
    March      May November  October September
        1        1        5        1        3
```

Although the months clearly have an ordering, this is not reflected in the output of the table function. Additionally, comparison operators are not supported for unordered factors. Creating an ordered factor solves these problems:

```
> mons = factor(mons,levels=c("January","February","March",
+                "April","May","June","July","August","September",
+                "October","November","December"),ordered=TRUE)
> mons[1] < mons[2]
[1] TRUE
```

```
> table(mons)
mons
   January  February     March     April       May      June
         3         2         1         2         1         0
      July    August September   October  November  December
         1         4         3         1         5         1
```

The order in which the levels are displayed is determined by the order in which they appear in the levels= argument to factor.

In the previous example, the levels of the factors had a natural ordering. Sometimes, a factor needs to be reordered on the basis of some property of that factor. For example, consider the InsectSpray data frame, which contains data on the numbers of insects seen (count) when an experimental unit was treated with one of six sprays (spray). The spray variable is stored as a factor with default ordering:

```
> levels(InsectSprays$spray)
[1] "A" "B" "C" "D" "E" "F"
```

Suppose we wish to reorder the factor levels of spray based on the mean value of the count variable for each level of spray. The reorder function takes three arguments: a factor, a vector of values on which the reordering is based, and a function to operate on those values for each factor level. Suppose we wish to reorder the levels of spray so that they are stored in the order of the mean value of count for each level of spray:

```
> InsectSprays$spray = with(InsectSprays,
+                           reorder(spray,count,mean))
> levels(InsectSprays$spray)
[1] "C" "E" "D" "A" "B" "F"
```

When reorder is used, it assigns an attribute called scores which contains the value used for the reordering:

```
> attr(InsectSprays$spray,'scores')
        A         B         C         D         E         F
14.500000 15.333333  2.083333  4.916667  3.500000 16.666667
```

As always, changes to system datasets are made in the local workspace; the original dataset is unchanged.

For some statistical procedures, the interpretation of results can be simplified by forcing a particular order to a factor; in particular, it may be useful to choose a "reference" level, which should be the first level of the factor. The relevel function allows you to choose a reference level, which will then be treated as the first level of the factor. For example, to make level "C" of InsectSprays$spray the first level, we could call relevel as follows:

```
> levels(InsectSprays$spray)
[1] "A" "B" "C" "D" "E" "F"
```

```
> InsectSprays$spray = relevel(InsectSprays$spray,'C')
> levels(InsectSprays$spray)
[1] "C" "A" "B" "D" "E" "F"
```

5.2 Numeric Factors

While it may be necessary to convert a numeric variable to a factor for a particular application, it is often very useful to convert the factor back to its original numeric values, since even simple arithmetic operations will fail when using factors. Since the as.numeric function will simply return the internal integer values of the factor, the conversion must be done using the levels attribute of the factor, or by first converting the factor to a character value using as.character.

Suppose we are studying the effects of several levels of a fertilizer on the growth of a plant. For some analyses, it might be useful to convert the fertilizer levels to an ordered factor:

```
> fert = c(10,20,20,50,10,20,10,50,20)
> fert = factor(fert,levels=c(10,20,50),ordered=TRUE)
> fert
[1] 10 20 20 50 10 20 10 50 20
Levels: 10 < 20 < 50
```

If we wished to calculate the mean of the original numeric values of the fert variable, we would have to convert the values using the levels function or as.character:

```
> mean(fert)
[1] NA
Warning message:
argument is not numeric or logical:
        returning NA in: mean.default(fert)
> mean(as.numeric(levels(fert)[fert]))
[1] 23.33333
> mean(as.numeric(as.character(fert)))
[1] 23.33333
```

Either method will achieve the desired result.

5.3 Manipulating Factors

When a factor is first created, all of its levels are stored along with the factor, and if subsets of the factor are extracted, they will retain all of the original levels. This can create problems when constructing model matrices and may or may not be useful when displaying the data using, say, the table function.

As an example, consider a random sample from the `letters` vector, which is part of the base R distribution:

```
> lets = sample(letters,size=100,replace=TRUE)
> lets = factor(lets)
> table(lets[1:5])


a b c d e f g h i j k l m n o p q r s t u v w x y z
0 0 1 0 0 0 0 1 0 1 0 1 0 0 0 0 0 0 0 0 0 0 0 1 1 0 0
```

Even though only five of the levels were actually represented, the `table` function shows the frequencies for all of the levels of the original factors. To change this, we can use the `drop=TRUE` argument to the subscripting operator. When used with factors, this argument will remove the unused levels:

```
> table(lets[1:5,drop=TRUE])


c h j w x
1 1 1 1 1
```

A similar result can be achieved by creating a new factor:

```
> table(factor(lets[1:5]))


c h j w x
1 1 1 1 1
```

To exclude certain levels from appearing in a factor, the `exclude=` argument can be passed to `factor`. By default, the missing value (`NA`) is excluded from factor levels; to create a factor that includes missing values from a numeric variable, use `exclude=NULL`.

Care must be taken when combining variables which are factors, because the `c` function will interpret the factors as integers. To combine factors, they should first be converted back to their original values (through the `levels` function), then catenated and converted to a new factor:

```
> fact1 = factor(sample(letters,size=10,replace=TRUE))
> fact2 = factor(sample(letters,size=10,replace=TRUE))
> fact1
 [1] o b i v q n q w e z
Levels: b e i n o q v w z
> fact2
 [1] b a s b l r g m z o
Levels: a b g l m o r s z
> fact12 = factor(c(levels(fact1)[fact1],
                    levels(fact2)[fact2]))
> fact12
 [1] o b i v q n q w e z b a s b l r g m z o
Levels: a b e g i l m n o q r s v w z
```

5.4 Creating Factors from Continuous Variables

The cut function is used to convert a numeric variable into a factor. The breaks= argument to cut is used to describe how ranges of numbers will be converted to factor values. If a number is provided through the breaks= argument, the resulting factor will be created by dividing the range of the variable into that number of equal-length intervals; if a vector of values is provided, the values in the vector are used to determine the breakpoints. Note that if a vector of values is provided, the number of levels of the resultant factor will be one less than the number of values in the vector.

For example, consider the women dataset, which contains height and weights for a sample of women. If we wanted to create a factor corresponding to weight, with three equally spaced levels, we could use the following:

```
> wfact = cut(women$weight,3)
> table(wfact)
wfact
(115,131] (131,148] (148,164]
        6         5         4
```

Notice that the default label for factors produced by cut contains the actual range of values that were used to divide the variable into factors. The pretty function can be used to choose cut points that are round numbers, but it may not return the number of levels that's actually desired:

```
> wfact = cut(women$weight,pretty(women$weight,3))
> wfact
 [1] (100,120] (100,120] (100,120] (120,140]
 [5] (120,140] (120,140] (120,140] (120,140]
 [9] (120,140] (140,160] (140,160] (140,160]
[13] (140,160] (140,160] (160,180]
4 Levels: (100,120] (120,140] (140,160] (160,180]
> table(wfact)
wfact
(100,120] (120,140] (140,160] (160,180]
        3         6         5         1
```

The labels= argument to cut allows you to specify the levels of the factors:

```
> wfact = cut(women$weight,3,labels=c('Low','Medium','High'))
> table(wfact)
wfact
   Low Medium   High
     6      5      4
```

To produce factors based on percentiles of your data (for example, quartiles or deciles), the quantile function can be used to generate the breaks= argument, insuring nearly equal numbers of observations in each of the levels of the factor:

```
> wfact = cut(women$weight,quantile(women$weight,(0:4)/4))
> table(wfact)
wfact
(115,124] (124,135] (135,148] (148,164]
        3         4         3         4
```

5.5 Factors Based on Dates and Times

As mentioned in Section 4.6, there are a number of ways to create factors from date/time objects. If you wish to create a factor based on one of the components of that date, you can extract it with strftime and convert it to a factor directly. For example, we can use the seq function to create a vector of dates representing each day of the year:

```
> everyday = seq(from=as.Date('2005-1-1'),
+               to=as.Date('2005-12-31'),by='day')
```

To create a factor based on the month of the year in which each date falls, we can extract the month name (full or abbreviated) using format:

```
> cmonth = format(everyday,'%b')
> months = factor(cmonth,levels=unique(cmonth),ordered=TRUE)
> table(months)
months
Jan Feb Mar Apr May Jun Jul Aug Sep Oct Nov Dec
 31  28  31  30  31  30  31  31  30  31  30  31
```

Since unique returns unique values in the order they are encountered, the levels argument will provide the month abbreviations in the correct order to produce a properly ordered factor.

For more details on formatting dates, see Section 4.3.

Sometimes more flexibility can be achieved by using the cut function, which understands time units of months, days, weeks, and years through the breaks= argument. (For date/time values, units of hours, minutes, and seconds can also be used.) For example, to format the days of the year based on the week in which they fall, we could use cut as follows:

```
> wks = cut(everyday,breaks='week')
> head(wks)
[1] 2004-12-27 2004-12-27 2005-01-03 2005-01-03
[5] 2005-01-03 2005-01-03
53 Levels: 2004-12-27 2005-01-03 ... 2005-12-26
```

Note that the first observation had a date earlier than any of the dates in the everyday vector, since the first date was in middle of the week. By default, cut starts weeks on Mondays; to use Sundays instead, pass the start.on.monday=FALSE argument to cut.

Multiples of units can also be specified through the `breaks=` argument. For example, to create a factor based on the quarter of the year an observation is in, we could use `cut` as follows:

```
> qtrs = cut(everyday,"3 months",labels=paste('Q',1:4,sep=''))
> head(qtrs)
[1] Q1 Q1 Q1 Q1 Q1 Q1
Levels: Q1 Q2 Q3 Q4
```

5.6 Interactions

Sometimes it is useful to treat all combinations of several factors as if they were a single factor. In situations like these, the `interaction` function can be used. This function will take two or more factors, and create a new, unordered factor whose levels correspond to the combinations of the levels of the input factors. For example, consider the data frame CO2, with factors `Plant`, `Type`, and `Treatment`. Suppose we wish to create a new factor representing the interaction of `Plant` and `Type`:

```
> data(CO2)
> newfact = interaction(CO2$Plant,CO2$Type)
> nlevels(newfact)
[1] 24
```

The factor `Plant` has 12 levels, and `Type` has two, resulting in 24 levels in the new factor. However, some of these combinations never occur in the dataset. Thus, `interaction`'s default behavior is to include all possible combinations of its input factors. To retain only those combinations for which there were observations, the `drop=TRUE` argument can be passed to `interaction`:

```
> newfact1 = interaction(CO2$Plant,CO2$Type,drop=TRUE)
> nlevels(newfact1)
[1] 12
```

By default, `interaction` forms levels for the new factor by joining the levels of its component factors with a period (`.`). This can be overridden with the `sep=` argument.

6

Subscripting

6.1 Basics of Subscripting

For objects that contain more than one element (vectors, matrices, arrays, data frames, and lists), subscripting is used to access some or all of those elements. Besides the usual numeric subscripts, R allows the use of character or logical values for subscripting. Subscripting operations are very fast and efficient, and are often the most powerful tool for accessing and manipulating data in R. The next subsections describe the different type of subscripts supported by R, and later sections will address the issues of using subscripts for particular data types.

6.2 Numeric Subscripts

Like most computer languages, numeric subscripts can be used to access the elements of a vector, array, or list. The first element of an object has subscript 1; subscripts of 0 are silently ignored. In addition to a single number, a vector of subscripts (or, for example, a function call that returns a vector of subscripts) can be used to access multiple elements. The colon operator and the `seq` function are especially useful here; see Section 2.8.1 for details.

Negative subscripts in R extract all of the elements of an object except the ones specified in the negative subscript; thus, when using numeric subscripts, subscripts must be either all positive (or zero) or all negative (or zero).

6.3 Character Subscripts

If a subscriptable object is named, a character string or vector of character strings can be used as a subscript. Negative character subscripts are not permitted; if you need to exclude elements based on their names, the `grep` function (Section 7.7) can be used. Like other forms of subscripting, a call to any function that returns a character string or vector of strings can be used as a subscript.

6.4 Logical Subscripts

Logical values can be used to selectively access elements of a subscriptable object, provided the size of the logical object is the same as the object (or part of the object) that is being subscripted. Elements corresponding to TRUE values in the logical vector will be included, and objects corresponding to FALSE values will not. Logical subscripting provides a very powerful and simple way to perform tasks that might otherwise require loops, while increasing the efficiency of your program as well. The first step in understanding logical subscripts is to examine the result of some logical expressions. Suppose we have a vector of numbers, and we're interested in those numbers which are more than 10. We can see where those numbers are with a simple logical expression.

```
> nums = c(12,9,8,14,7,16,3,2,9)
> nums > 10
[1]   TRUE FALSE FALSE   TRUE FALSE   TRUE FALSE FALSE FALSE
```

Like most operations in R, logical operators are vectorized; applying a logical subscript to a vector or an array will produce an object of the same size and shape as the original object. In this example, we applied a logical operation to a vector of length 10, and it returned a logical vector of length 10, with TRUE in each position where the value in the original vector was greater than 10, and FALSE elsewhere. If we use this logical vector for subscripting, it will extract the elements for which the logical vector is true:

```
> nums[nums>10]
[1] 12 14 16
```

For the closely related problem of finding the indices of these elements, R provides the which function, which accepts a logical vector, and returns a vector containing the subscripts of the elements for which the logical vector was true:

```
> which(nums>10)
[1] 1 4 6
```

In this simple example, the operation is the equivalent of

```
> seq(along=nums)[nums > 10]
[1] 1 4 6
```

Logical subscripts allow for modification of elements that meet a particular condition by using an appropriately subscripted object on the left-hand side of an assignment statement. If we wanted to change the numbers in nums that were greater than 10 to zero, we could use

```
> nums[nums > 10] = 0
> nums
[1] 0 9 8 0 7 0 3 2 9
```

6.5 Subscripting Matrices and Arrays

Multidimensional objects like matrices introduce a new type of subscripting: the empty subscript. For a multidimensional object, subscripts can be provided for each dimension, separated by commas. For example, we would refer to the element of a matrix x in the fourth row and third column as x[4,3]. If we omit, say, the second subscript and refer to x[4,], the subscripting operation will apply to the entire dimension that was omitted; in this case, all of the columns in the fourth row of x. Thus, accessing entire rows and columns is simple; just leave out the subscript for the dimension you're not interested in. The following examples show how this can be used:

```
> x = matrix(1:12,4,3)
> x
     [,1] [,2] [,3]
[1,]    1    5    9
[2,]    2    6   10
[3,]    3    7   11
[4,]    4    8   12
> x[,1]
[1] 1 2 3 4
> x[,c(3,1)]
     [,1] [,2]
[1,]    9    1
[2,]   10    2
[3,]   11    3
[4,]   12    4
> x[2,]
[1]  2  6 10
> x[10]
[1] 10
```

Pay careful attention to the last example, where a matrix is subscripted with a single subscript. In this case, the matrix is silently treated like a vector composed of all the columns of the matrix. While this may be useful in certain situations, you should generally use two subscripts when working with matrices.

Notice that by manipulating the order of subscripts, we can create a sub-matrix with rows or columns in whatever order we want. This fact coupled with the order function provides a method to sort a matrix or data frame in the order of any of its columns. The order function returns a vector of indices that will permute its input argument into sorted order. Perhaps the best way to understand order is to consider that x[order(x)] will always be identical to sort(x). Suppose we wish to sort the rows of the stack.x matrix by increasing values of the Air.Flow variable. We can use order as follows:

```
> stack.x.a = stack.x[order(stack.x[,'Air.Flow']),]
> head(stack.x.a)
    Air.Flow Water.Temp Acid.Conc.
15        50         18         89
16        50         18         86
17        50         19         72
18        50         19         79
19        50         20         80
20        56         20         82
```

Note the comma after the call to order, indicating that we wish to rearrange all the columns of the matrix in the order of the specified variable. To reverse the order of the resulting sort, use the decreasing=TRUE argument to order. Although the order function accepts multiple arguments to allow ordering by multiple variables, it is sometimes inconvenient to have to list each such argument in the function call. For example, we might want a function which can accept a variable number of ordering variables, and which will then call order properly, regardless of how many arguments are used. Problems like this can be easily handled in R with the do.call function. The idea behind do.call is that it takes a list of arguments and prepares a call to a function of your choice, using the list elements as if they had been passed to the function as individual arguments. The first argument to do.call is a function or a character variable containing the name of a function, and the only other required argument is a list containing the arguments that should be passed to the function. Using do.call, we can write a function to sort the rows of a data frame by any number of its columns:

```
sortframe = function(df,...)df[do.call(order,list(...)),]
```

(When used inside a function allowing multiple unnamed arguments, the ex-pression list(...) creates a list containing all the unnamed arguments.) For example, to sort the rows of the iris data frame by Sepal.Length and Sepal.Width, we could call sortframe as follows:

```
> with(iris,sortframe(iris,Sepal.Length,Sepal.Width))
      Sepal.Length Sepal.Width Petal.Length Petal.Width     Species
14            4.3         3.0          1.1         0.1       setosa
9             4.4         2.9          1.4         0.2       setosa
39            4.4         3.0          1.3         0.2       setosa
43            4.4         3.2          1.3         0.2       setosa
42            4.5         2.3          1.3         0.3       setosa
4             4.6         3.1          1.5         0.2       setosa
48            4.6         3.2          1.4         0.2       setosa
7             4.6         3.4          1.4         0.3       setosa
23            4.6         3.6          1.0         0.2       setosa
                             . . .
```

Another common operation, reversing the order of rows or columns of a matrix , can be achieved through the use of a call to the rev function as either the row or column subscript. For example, to create a version of the iris data frame whose rows are in the reverse order of the original, we could use

```
> riris = iris[rev(1:nrow(iris)),]
> head(riris)
      Sepal.Length Sepal.Width Petal.Length Petal.Width     Species
150           5.9         3.0          5.1         1.8    virginica
149           6.2         3.4          5.4         2.3    virginica
148           6.5         3.0          5.2         2.0    virginica
147           6.3         2.5          5.0         1.9    virginica
146           6.7         3.0          5.2         2.3    virginica
145           6.7         3.3          5.7         2.5    virginica
```

By default, subscripting operations reduce the dimensions of an array whenever possible. The result of this is that functions will sometimes fail when passed a single row or column from a matrix, since subscripting can potentially return a vector, even though the subscripted object is an array. To prevent this from happening the array nature of the extracted part can be retained with the drop=FALSE argument, which is passed along with the subscripts of the array. This example shows the effect of using this argument:

```
> x = matrix(1:12,4,3)
> x[,1]
[1] 1 2 3 4
> x[,1,drop=FALSE]
      [,1]
[1,]    1
[2,]    2
[3,]    3
[4,]    4
```

Note the "extra" comma inside the subscripting brackets – `drop=FALSE` is considered an argument to the subscripting operation. `drop=FALSE` may also prove useful if a named column loses its name when passed to a function.

Using subscripts, it's easy to selectively access any combination of rows and/or columns that you need. Suppose we want to find all of the rows in x for which the first column is less than 3. Since we want all the elements of these rows, we will use an empty subscript for the column (second) dimension. Once again it may be instructive to examine the subscript used for the first dimension:

```
> x[,1] < 3
[1]   TRUE   TRUE FALSE FALSE
> x[x[,1] < 3,]
     [,1] [,2] [,3]
[1,]   1    5    9
[2,]   2    6   10
```

The logical vector `x[,1] < 3` is of length 4, the number of rows in the matrix; thus, it can be used as a logical subscript for the first dimension to specify the rows we're interested in. By using the expression with an empty second subscript, we extract all of the columns for these rows.

Matrices allow an additional special form of subscripting. If a two-column matrix is used as a subscript for a matrix, the elements specified by the row and column combination of each line will be accessed. This makes it easy to create matrices from tabular values. Consider the following matrix, whose first two columns represent a row and column number, and whose last column represents a value:

```
> mat = matrix(scan(),ncol=3,byrow=TRUE)
1: 1 1 12 1 2 7 2 1 9 2 2 16 3 1 12 3 2 15
19:
Read 18 items
> mat
     [,1] [,2] [,3]
[1,]   1    1   12
[2,]   1    2    7
[3,]   2    1    9
[4,]   2    2   16
[5,]   3    1   12
[6,]   3    2   15
```

The row and column numbers found in the first two columns describe a matrix with three rows and two columns; we first create a matrix of missing values to hold the result, and then use the first two columns of the matrix as the subscript, with the third column being assigned to the new matrix:

```
> newmat = matrix(NA,3,2)
> newmat[mat[,1:2]] = mat[,3]
```

```
> newmat
      [,1] [,2]
[1,]   12    7
[2,]    9   16
[3,]   12   15
```

Any elements whose values were not specified will retain their original values, in this case a value of NA. See the discussion of xtabs in Section 8.1 for an alternative method of converting tabulated data into an R table.

6.6 Specialized Functions for Matrices

Two simple functions, while not very useful on their own, extend the power of subscripting for matrices based on the relative positions of matrix elements. The row function, when passed a matrix, returns a matrix of the identical dimensions with the row numbers of each element, while col plays the same role, but uses the column numbers. For example, consider an artificial contingency table showing the results of two different classification methods for a set of objects:

```
> method1 = c(1,1,1,1,2,2,2,2,3,3,3,3,4,4,4,4)
> method2 = c(1,2,2,3,2,2,1,3,3,3,2,4,1,4,4,3)
> tt = table(method1,method2)
> tt
       method2
method1 1 2 3 4
      1 1 2 1 0
      2 1 2 1 0
      3 0 1 2 1
      4 1 0 1 2
```

Suppose we want to extract all the off-diagonal elements. One way to think about these elements is that their row number and column numbers are different. Expressed using the row and col functions, this is equivalent to

```
> offd = row(tt) != col(tt)
> offd
       [,1]  [,2]  [,3]  [,4]
[1,] FALSE  TRUE  TRUE  TRUE
[2,]  TRUE FALSE  TRUE  TRUE
[3,]  TRUE  TRUE FALSE  TRUE
[4,]  TRUE  TRUE  TRUE FALSE
```

Since this matrix is the same size as tt, it can be used as a subscript to extract the off-diagonal elements:

```
> tt[offd]
 [1] 1 0 1 2 1 0 1 1 1 0 0 1
```

So, for example, we could calculate the sum of the off-diagonal elements as

```
> sum(tt[offd])
```

The R functions `lower.tri` and `upper.tri` use this technique to return a logical matrix useful in extracting the lower or upper triangular elements of a matrix. Each accepts a `diag=` argument; setting this argument to `TRUE` will set the diagonal elements of the matrix to `TRUE` along with the off-diagonal ones.

The `diag` function can be used to extract or set the diagonal elements of a matrix, or to form a matrix which has specified values on the diagonals.

6.7 Lists

Lists are the most general way to store a collection of objects in R, because there is no limitation on the mode of the objects that a list may hold. Although it hasn't been explicitly stated, one rule of subscripting in R is that subscripting will always return an object of the same mode as the object being subscripted. For matrices and vectors, this is completely natural, and should never cause confusion. But for lists, there is a subtle distinction between part of a list, and the object which that part of the list represents. As a simple example, consider a list with some names and some numbers:

```
> simple = list(a=c('fred','sam','harry'),b=c(24,17,19,22))
> mode(simple)
[1] "list"
> simple[2]
$b
[1]  24 17 19 22

> mode(simple[2])
[1] "list"
```

Although it looks as if `simple[2]` represents the vector, it's actually a list containing the vector; operations that would work on the vector will fail on this list:

```
> mean(simple[2])
[1] NA
Warning message:
argument is not numeric or logical:
        returning NA in: mean.default(simple[2])
```

R provides two convenient ways to resolve this issue. First, if the elements of the list are named, the actual contents of the elements can be accessed by separating the name of the list from the name of the element with a dollar sign ($). So we could get around the previous problem by referring to `simple[2]` as `simple$b`. For interactive sessions, using the dollar sign notation is the natural way to perform operations on the elements of a list.

For those situations where the dollar sign notation would be inappropriate (for example, accessing elements through their index or through a name stored in a character variable), R provides the double bracket subscript operator. Double brackets are not restricted to respect the mode of the object they are subscripting, and will extract the actual list element from the list. So in order to find the mean of a numeric list element we could use any of these three forms:

```
> mean(simple$b)
[1] 20.5
> mean(simple[[2]])
[1] 20.5
> mean(simple[['b']])
[1] 20.5
```

The key thing to notice is that in this case, single brackets will always return a list containing the selected element(s), while double brackets will return the actual contents of selected list element. This difference can be visualized by printing the two different forms:

```
> simple[1]
$a
[1] "fred"   "sam"    "harry"

> simple[[1]]
[1] "fred"   "sam"    "harry"
```

The "`$a`" is an indication that the object being displayed is a list, with a single element named a, not a vector. Notice that double brackets are not appropriate for ranges of list elements; in these cases single brackets must be used. For example, to access both elements of the `simple` list, we could use `simple[c(1,2)]`, `simple[1:2]`, or `simple[c('a','b')]`, but using `simple[[1:2]]` would not produce the expected result.

6.8 Subscripting Data Frames

Since data frames are a cross between a list and a matrix, it's not surprising that both matrix and list subscripting techniques apply to data frames. One of the few differences regards the use of a single subscript; when a single subscript is used with a data frame, it behaves like a list rather than a vector, and the subscripts refer to the columns of the data frame, which are its list elements.

When using logical subscripts with data frames containing missing values, it may be necessary to remove the missing values before the logical comparison

is made, or unexpected results may occur. For example, consider this small
data frame where we want to find all the rows where b is greater than 10:

```
> dd = data.frame(a=c(5,9,12,15,17,11),b=c(8,NA,12,10,NA,15))
> dd[dd$b > 10,]
      a  b
NA   NA NA
3    12 12
NA.1 NA NA
6    11 15
```

Along with the desired results are additional rows wherever a missing value
appeared in b. The problem is easily remedied by using a more complex logical
expression that insures missing values will generate a value of FALSE instead
of NA:

```
> dd[!is.na(dd$b) & dd$b > 10,]
   a  b
3 12 12
6 11 15
```

This situation is so common that R provides the subset function which accepts a data frame, matrix or vector, and a logical expression as its first two
arguments, and which returns a similar object containing only those elements
that meet the condition of the logical expression. It insures that missing values
don't get included, and, if its first argument is a data frame or matrix with
named columns, it also resolves variable names inside the logical expression
from the object passed as the first argument. So subset could be used for the
previous example as follows:

```
> subset(dd,b>10)
   a  b
3 12 12
6 11 15
```

Notice that it's not necessary to use the data frame name when referring
to variables in the subsetting argument. A further convenience is offered by
the select= argument which will extract only the specified columns from
the data frame passed as the first argument. The argument to select= is a
vector of integers or variable names which correspond to the columns that are
to be extracted. Unlike most other functions in R, names passed through the
select= argument can be either quoted or unquoted. To ignore columns, their
name or index number can be preceded by a negative sign (-). For example,
consider the LifeCycleSavings data frame distributed with R. Suppose we
want to create a data frame containing the variables pop15 and pop75 for
those observations in the data frame for which sr is greater than 10. The
following expression will create the data frame:

```
> some = subset(LifeCycleSavings,sr>10,select=c(pop15,pop75))
```

Since the `select=` argument works by replacing variable names with their corresponding column indices, ranges of columns can be specified using variable names:

```
> life1 = subset(LifeCycleSavings,select=pop15:dpi)
```

will extract columns starting at `pop15` and ending at `dpi`. Since these are the first three columns of the data frame, an equivalent specification would be

```
> life1 = subset(LifeCycleSavings,select=1:3)
```

Similarly, we could create a data frame like `LifeCycleSavings`, but without the `pop15` and `pop75` columns with expressions like the following:

```
> life2 = subset(LifeCycleSavings,select=c(-pop15,-pop75))
```

or

```
> life2 = subset(LifeCycleSavings,select=-c(2,3))
```

Remember that the `subset` function will always return a new data frame, matrix or vector, so it is not suited for modifying selected parts of a data frame. In those cases, the basic subscripting operations described above must be used.

7

Character Manipulation

While R is usually thought of as a language designed for numerical computation, it contains a full complement of functions which can manipulate character data. Combined with R's powerful vectorized operations, these functions can perform the same sorts of tasks that scripting languages like perl and python are often used for.

7.1 Basics of Character Data

Character values in R can be stored as scalars, vectors, or matrices, or they can be columns of a data frame or elements of a list. When applied to objects like this, the `length` function will report the number of character values in the object, not the number of characters in each string. To find the number of characters in a character value, the `nchar` function can be used. Like most functions in R, `nchar` is vectorized. For example, the names of the fifty states in the United States can be found in the vector `state.name` which is distributed as part of R. To find the lengths of the names of the states, `nchar` can be used:

```
> nchar(state.name)
 [1]  7  6  7  8 10  8 11  8  7  7  6  5  8  7  4
[16]  6  8  9  5  8 13  8  9 11  8  7  8  6 13 10
[31] 10  8 14 12  4  8  6 12 12 14 12  9  5  4  7
[46]  8 10 13  9  7
```

7.2 Displaying and Concatenating Character Strings

Like other objects in R, character values will be displayed when their name is typed at the console or when they are passed to the `print` function. However, it is often more convenient to print or display these objects directly without the subscripts that the print function provides. The cat function will combine character values and print them to the screen or a file directly. The cat function coerces its arguments to character values, then concatenates and displays them. This makes the function ideal for printing messages and warnings from inside of functions:

```
> x = 7
> y = 10
> cat('x should be greater than y, but x=',x,'and y=',y,'\n')
x should be greater than y, but x= 7 and y= 10
```

Note the use of a newline (\n) in the argument list to insure that a complete line is displayed. cat will always print a newline when it encounters a newline character. When there are multiple strings passed to cat, or when the argument to cat is a vector of character strings, the fill= argument can be used to automatically insert newlines into the output string. If fill= is set to TRUE, the value of the system width option will be used to determine the linesize; if a numeric value is used, the output will be displayed using that width, although cat will not insert newlines into individual elements of its input:

```
> cat('Long strings can','be displayed over',
+     'several lines using','the fill= argument',
+     fill=40)
Long strings can be displayed over
several lines using the fill= argument
```

The cat function also accepts a file= argument to specify that its output should be directed to a file. When the file= argument is used, the append=TRUE argument can also be provided to have cat append its output to an already existing file.

For more control over the way that character values are concatenated, the paste function can be used. In its simplest usage, this function will accept an unlimited number of scalars, and join them together, separating each scalar with a space by default. To use a character string other than a space as a separator, the sep= argument can be used. If any object passed to paste is not of mode character, it is converted to character:

```
> paste('one',2,'three',4,'five')
[1] "one 2 three 4 five"
```

If a character vector is passed to paste, the collapse= argument can be used to specify a character string to place between each element of the vector:

```
> paste(c('one','two','three','four'),collapse=' ')
[1] "one two three four"
```

Note that the collapse= argument must be used in these cases, as sep= has no effect when applied to a vector.

When multiple arguments are passed to `paste`, it will vectorize the operation, recycling shorter elements when necessary. This makes it easy to generate variable names with a common prefix:

```
> paste('X',1:5,sep='')
[1] "X1" "X2" "X3" "X4" "X5"
> paste(c('X','Y'),1:5,sep='')
[1] "X1" "Y2" "X3" "Y4" "X5"
```

In cases like this, the `sep=` argument controls what is placed between each set of values that are combined, and the `collapse=` argument can be used to specify a value to use when joining those individual values to create a single string:

```
> paste(c('X','Y'),1:5,sep='_',collapse='|')
[1] "X_1|Y_2|X_3|Y_4|X_5"
```

The same sort of operations can be applied to multiple arguments to `paste`:

```
> paste(c('X','Y'),1:5,'^',c('a','b'),sep='_',collapse='|')
[1] "X_1_^_a|Y_2_^_b|X_3_^_a|Y_4_^_b|X_5_^_a"
```

By omitting the `collapse` argument, the individual pasted pieces are returned separately instead of being joined into a single string:

```
> paste(c('X','Y'),1:5,'^',c('a','b'),sep='_')
[1] "X_1_^_a" "Y_2_^_b" "X_3_^_a" "Y_4_^_b" "X_5_^_a"
```

7.3 Working with Parts of Character Values

Individual characters of character values are not accessible through ordinary subscripting. Instead, the `substring` function can be used either to extract parts of character strings, or to change the values of parts of character strings. In addition to the string being operated on, `substring` accepts a `first=` argument giving the first character of the desired substring, and a `last=` argument giving the last character. If not specified, `last=` defaults to a large number, so that specifying just a `start=` value will operate from that character to the end of the string. Like most functions in R, `substring` is vectorized, operating on multiple strings at once:

```
> substring(state.name,2,6)
 [1] "labam" "laska" "rizon" "rkans" "alifo" "olora" "onnec"
 [8] "elawa" "lorid" "eorgi" "awaii" "daho"  "llino" "ndian"
[15] "owa"   "ansas" "entuc" "ouisi" "aine"  "aryla" "assac"
[22] "ichig" "innes" "issis" "issou" "ontan" "ebras" "evada"
[29] "ew Ha" "ew Je" "ew Me" "ew Yo" "orth " "orth " "hio"
[36] "klaho" "regon" "ennsy" "hode " "outh " "outh " "ennes"
[43] "exas"  "tah"   "ermon" "irgin" "ashin" "est V" "iscon"
[50] "yomin"
```

Notice that in the case of strings that have fewer characters than specified in the `last=` argument (like `Ohio` or `Texas` in this example), `substring` returns as many characters as it finds with no padding provided. (The `sprintf` function can be used to pad a series of character values to a common size; see Section 2.13.)

Vectorization takes place for the `first=` and `last=` arguments as well as for character vectors passed to `subscript`. Although the `strsplit` function described in Section 7.6 can perform the operation automatically, a vector of character values can be created from a single string by `substring` as follows:

```
> mystring = 'dog cat duck'
> substring(mystring,c(1,5,9),c(3,7,12))
[1] "dog"  "cat"  "duck"
```

For finding locations of particular characters within a character string, the string first needs to be converted to a character vector containing individual characters. This can be done by passing a vector consisting of all the characters to be processed as both the `first=` and `last=` arguments, and then applying `which` to the result:

```
> state = 'Mississippi'
> ll = nchar(state)
> ltrs = substring(state,1:ll,1:ll)
> ltrs
 [1] "M" "i" "s" "s" "i" "s" "s" "i" "p" "p" "i"
> which(ltrs == 's')
[1] 3 4 6 7
```

The assignment form of `substring` allows replacement of selected portions of character strings, but `substring` will only replace parts of the string with values that have the same number of characters; if a string that's shorter than the implied substring is provided, none of the original string will be overwritten:

```
> mystring = 'dog cat duck'
> substring(mystring,5,7) = 'feline'
> mystring
[1] "dog fel duck"
> mystring = 'dog cat duck'
> substring(mystring,5,7) = 'a'
> mystring
[1] "dog aat duck"
```

7.4 Regular Expressions in R

Regular expressions are a method of expressing patterns in character values which can then be used to extract parts of strings or to modify those strings in some way. Regular expressions are supported in the R functions `strsplit`, `grep`, `sub`, and `gsub`, as well as in the `regexpr` and `gregexpr` functions which are the main tools for working with regular expressions in R.

Regular expression syntax varies depending on the particular implementation a program uses. R tries to provide a great deal of flexibility regarding the regular expressions it understands. By default, R uses a basic set of regular expressions similar to those used by UNIX utilities like `grep`. The `extended=TRUE` argument to R functions that support regular expressions extend the set of regular expressions to include those supported by the POSIX 1003.2 standard. To use regular expressions like those supported by scripting languages such as perl and python, the `perl=TRUE` argument can be used. Thus, if you're already familiar with a particular type of regular expressions, you can probably find an option that will make R work the way you expect it to.

The backslash character (\) is used in regular expressions to signal that certain characters with special meaning in regular expressions should be treated as normal characters. In R, this means that two backslash characters need to be entered into an input string anywhere that special characters need to be escaped. Although the double backslash will display when the string is printed, `nchar` or `cat` can verify that only a single backslash is actually included in the string. For example, in regular expressions, a period (.) is ordinarily matched by any single character. To create a regular expression that would match file names with an extension of ".txt", we could use a regular expression like

```
> expr = '.*\\.txt'
> nchar(expr)
[1] 7
> cat(expr,'\n')
.*\.txt
```

Single backslashes, like those which are part of a newline character (\n), will be interpreted correctly inside of regular expressions. One way to avoid the need for quotes or double backslashes is to use the `readline` function to enter your regular expressions into R. For example, we could use `readline` in the previous example as follows:

```
> expr = readline()
.*\.txt
> nchar(expr)
[1] 7
```

7.5 Basics of Regular Expressions

Regular expressions are composed of three components: literal characters, which are matched by a single character; character classes, which can be matched by any of a number of characters, and modifiers, which operate on literal characters or character classes. Since many punctuation marks are regular expression modifiers, the following characters must always be preceded by a backslash to retain their literal meaning:

. ^ $ + ? * ( ) [ ] { } | \

To form a character class, use square brackets ([]) surrounding the characters that you would like to match. For example, to create a character class that will be matched by either a, b, or 3, use [ab3]. Dashes can be used inside of character classes to represent a range of values such as [a-z] or [5-9]. Because of this, if a dash is to be literally included in a character class, it should either be the first character in the class or it should be preceded by a backslash. Other special characters (except square brackets) do not need a backslash when used in a character class.

Using characters and character classes as basic building blocks, we can now construct regular expressions by understanding the modifiers that are part of the regular expression language. These operators are listed in Table 7.1.

Modifier	Meaning
^	anchors expression to beginning of target
$	anchors expression to end of target
.	matches any single character except newline
		separates alternative patterns
()	groups patterns together
*	matches 0 or more occurrences of preceding entity
?	matches 0 or 1 occurrences of preceding entity
+	matches 1 or more occurrences of preceding entity
$\{n\}$	matches exactly n occurrences of preceding entity
$\{n,\}$	matches at least n occurrences of preceding entity
$\{n,m\}$	matches between n and m occurrences

Table 7.1. Modifiers for regular expressions

The modifiers operate on whatever entity they follow, using parentheses for grouping if necessary. As some simple examples, a string with two digits followed by one or more letters could be matched by the regular expression "[0-9][0-9][a-zA-Z]+"; three consecutive occurrences of the string "abc" could be matched by "(abc){3}"; a filename consisting of all letters, and ending in ".jpg" could be matched by "^[a-zA-Z]+\\.jpg$". (In the previous example, the double backslashes would be required if you entered the regular expression as a quoted string in R; if you used readline to enter the expression, you would use just a single backslash.)

Remember that regular expressions are simply character strings in R, so they can be manipulated like any other character strings. For example, the

vertical bar (|) is used in regular expressions to express alternation. To create
a regular expression that would be matched by several different strings, we
can combine the strings using the bar as a separator:

```
> strs = c('chicken','dog','cat')
> expr = paste(strs,collapse='|')
> expr
[1] "chicken|dog|cat"
```

The variable `expr` could now be used as a regular expression to match any of
the words in the original vector.

7.6 Breaking Apart Character Values

The `strsplit` function can use a character string or regular expression to divide up a character string into smaller pieces. The first argument to `strsplit`
is the character string to break up, and the second argument is the character
value or regular expression which should be used to break up the string into
parts.

Like other functions that can return different numbers of elements from
their inputs, `strsplit` returns its results as a list, even when its input is a
single character string. For example, suppose we want to break up a simple
sentence into individual words, by splitting the string wherever a blank occurs:

```
> sentence =
+ 'R is a free software environment for statistical computing'
> parts = strsplit(sentence,' ')
> parts
[[1]]
[1] "R"          "is"          "a"     "free"
[5] "software"   "environment" "for"   "statistical"
[9] "computing"
```

To access the results, the first element of the list must be used:

```
> length(parts)
[1] 1
> length(parts[[1]])
[1] 9
```

When the input to `strsplit` is a vector of character strings, `sapply` can be
used to process the output to return results for each of the strings:

```
> more = c('R is a free software environment for statistical
+           computing', 'It compiles and runs on a wide
            variety of UNIX platforms')
> result = strsplit(more,' ')
> sapply(result,length)
[1]  9 11
```

Alternatively, if the structure of the output is not important, all of the split parts can be combined using `unlist`:

```
> allparts = unlist(result)
> allparts
 [1] "R"         "is"          "a"         "free"
 [5] "software"  "environment" "for"       "statistical"
 [9] "computing" "It"          "compiles"  "and"
[13] "runs"      "on"          "a"         "wide"
[17] "variety"   "of"          "UNIX"      "platforms"
```

Because `strsplit` can accept regular expressions to decide where to split a character string, a wide variety of situations can be easily handled. For example, if there are multiple spaces in a string, and a space is used as the splitting character, extra empty strings may be returned:

```
> str = 'one  two   three four'
> strsplit(str,' ')
[[1]]
[1] "one"   ""       "two"   ""      ""      "three" "four"
```

By using a regular expression representing one or more blanks (using the + modifier), we can extract only nonempty strings:

```
> strsplit(str,' +')
[[1]]
[1] "one"   "two"   "three" "four"
```

Using an empty string as the splitting character, `strsplit` can return a list of individual characters from a vector of character strings:

```
> words = c('one two','three four')
> strsplit(words,'')
[[1]]
[1] "o" "n" "e" " " "t" "w" "o"

[[2]]
[1] "t" "h" "r" "e" "e" " " "f" "o" "u" "r"
```

7.7 Using Regular Expressions in R

The `grep` function accepts a regular expression and a character string or vector of character strings, and returns the indices of those elements of the strings which are matched by the regular expression. If the `value=TRUE` argument is passed to `grep`, it will return the actual strings which matched the expression instead of the indices.

If the string to be matched should be interpreted literally (i.e., not as a regular expression), the `fixed=TRUE` argument should be used.

One important use of grep is to extract a set of variables from a data frame based on their names. For example, the LifeCycleSavings data frame contains two variables with information about the percentage of population less than 15 years old (pop15) or greater than 75 years old (pop75). Since both of these variables begin with the string "pop", we can find their indices or values using grep:

```
> grep('^pop',names(LifeCycleSavings))
[1] 2 3
> grep('^pop',names(LifeCycleSavings),value=TRUE)
[1] "pop15" "pop75"
```

To create a data frame with just these variables, we can use the output of grep as a subscript:

```
> head(LifeCycleSavings[,grep('^pop',names(LifeCycleSavings))])
          pop15 pop75
Australia 29.35  2.87
Austria   23.32  4.41
Belgium   23.80  4.43
Bolivia   41.89  1.67
Brazil    42.19  0.83
Canada    31.72  2.85
```

To find regular expressions without regard to the case (upper or lower) of the input, the ignore.case=TRUE argument can be used. To search for the string "dog" appearing as a word and ignoring case, we could use the following:

```
> inp = c('run dog run','work doggedly','CAT AND DOG')
> grep('\\<dog\\>',inp,ignore.case=TRUE)
[1] 1 3
```

Surrounding a string with escaped angle brackets (\\< and \\>) restricts matches to the case where the string is surrounded by either white space, punctuation, or a line ending or beginning.

If the regular expression passed to grep is not matched in any of its inputs, grep returns an empty numeric vector. Thus, the any function can be used to test if a regular expression occurs anywhere in a vector of strings:

```
> str1 = c('The R Foundation','is a not for profit
+           organization','working in the public interest')
> str2 = c(' It was founded by the members',
+          'of the R Core Team in order',
+          'to provide support for the R project')
> any(grep('profit',str1))
[1] TRUE
> any(grep('profit',str2))
[1] FALSE
```

While the `grep` function can be used to test for the presence of a regular expression, sometimes more details regarding the matches that are found are needed. In R, the `regexpr` and `gregexpr` functions can be used to pinpoint and possibly extract those parts of a string that were matched by a regular expression. The output from these functions is a vector of starting positions of the regular expressions which were found; if no match occurred, a value of −1 is returned. In addition, an attribute called `match.length` is associated with the vector of starting positions to provide information about exactly which characters were involved in the match. The `regexpr` function will only provide information about the first match in its input string(s), while the `gregexpr` function returns information about all matches found. The input arguments to `regexpr` and `gregexpr` are similar to those of `grep`; however, the `ignore.case=TRUE` argument is not available in versions of R earlier than version 2.6.

Since `regexpr` only reports the first match it finds, it will always return a vector, with −1 in those positions where no match was found. To extract the strings that actually matched, `substr` can be used, after calculating the ending position from the `regexpr` output and the `match.length` attribute:

```
> tst = c('one x7 two b1','three c5 four b9',
+          'five six seven','a8 eight nine')
> wh = regexpr('[a-z][0-9]',tst)
> wh
[1]  5  7 -1  1
attr(,"match.length")
[1]  2  2 -1  2
> res = substring(tst,wh,wh + attr(wh,'match.length') - 1)
> res
[1] "x7" "c5" ""   "a8"
```

In the case of the third string, which did not contain the regular expression, an empty string is returned, preserving the structure of the output relative to the input. If empty strings are not desired, they can be easily removed:

```
> res[res != '']
[1] "x7" "c5" "a8"
```

The output from `gregexpr` is similar to that of `regexpr`, but, like `strsplit`, `gregexpr` always returns its result in the form of a list. Continuing the previous example, but looking for all matches, we can call `gregexpr` as follows:

```
> wh1 = gregexpr('[a-z][0-9]',tst)
> wh1
[[1]]
[1]  5 12
attr(,"match.length")
[1] 2 2
```

```
[[2]]
[1]   7 15
attr(,"match.length")
[1] 2 2

[[3]]
[1] -1
attr(,"match.length")
[1] -1

[[4]]
[1] 1
attr(,"match.length")
[1] 2
```

To further process the results from gregexpr, we need to call the substring function for each element of the output list. One way to do it is with a loop:

```
> res1 = list()
> for(i in 1:length(wh1))
+               res1[[i]] = substring(tst[i],wh1[[i]],
+                           wh1[[i]] +
+                           attr(wh1[[i]],'match.length') -1)
> res1
[[1]]
[1] "x7" "b1"

[[2]]
[1] "c5" "b9"

[[3]]
[1] ""


[[4]]
[1] "a8"
```

Another possibility for processing the output is to use mapply. The first argument to mapply is a function that accepts multiple arguments; the remaining arguments are vectors of equal lengths (like the text input and the output from gregexpr), whose elements will be passed to that function one at a time. The same technique used in the previous example can be encapsulated into a function as follows:

```
> getexpr = function(str,greg)substring(str,greg,
+                           greg + attr(greg,'match.length') - 1)
```

Now mapply can be called with the two vectors of interest:

```
> res2 = mapply(getexpr,tst,wh1)
> res2
$"one x7 two b1"
[1] "x7" "b1"

$"three c5 four b9"
[1] "c5" "b9"

$"five six seven"
[1] ""

$"a8 eight nine"
[1] "a8"
```

One advantage of this approach is that it automatically creates an appropriate object to hold the output; in addition, mapply uses the input strings as names in the output, which may or may not be desirable.

7.8 Substitutions and Tagging

For substituting text based on regular expressions, R provides two functions: sub and gsub. Each of these functions accepts a regular expression, a string containing what will be substituted for the regular expression, and the string or strings to operate on. The sub function changes only the first occurrence of the regular expression, while the gsub function performs the substitution on all occurrences within the string.

One important use of these functions concerns numeric data which is read from text sources like web pages or financial reports, and which may contain commas or dollar signs. For example, suppose we've input a vector of values from a financial report as follows:

```
> values = c('$11,317.35','$11,234.51','$11,275.89',
+                '$11,278.93','$11,294.94')
```

To use these values as numbers, the commas and dollar signs need to be removed before as.numeric can be used. A regular expression to find either commas or dollar signs can be composed using a character class, and this can be passed to gsub with an empty substitution pattern, providing values which can be converted to numbers:

```
> as.numeric(gsub('[$,]','',values))
[1] 11317.35 11234.51 11275.89 11278.93 11294.94
```

When using this technique, avoid using as.numeric on anything less than an entire vector of values, since the mode of an individual element of a matrix or a single value in a data frame cannot be changed.

When using the substitution functions, a powerful feature known as tagging of regular expressions is available. When part of a regular expression is surrounded by (unescaped) parentheses, that part can be used in a substitution pattern by representing it as a backslash followed by a number. The first tagged pattern is represented by \\1, the second by \\2, and so on. A common practice in financial reports is to surround values that represent negative numbers with parentheses; these parentheses will prevent R from properly interpreting such values as numbers. We can tag the number inside the parentheses using a regular expression, and substitute the value by preceding it with a minus sign. Note the difference between the literal parentheses (preceded by two backslashes) and the parentheses used for tagging:

```
> values = c('75.99','(20.30)','55.20')
> as.numeric(gsub('\\(([0-9.]+)\\)','-\\1',values))
[1]  75.99 -20.30  55.20
```

To extract just the tagged pattern from a regular expression, one possibility is to use the regular expression beginning and end anchor characters (^ and $, respectively) to account for all the nontagged characters in the string, and specify just the tagged expression for the substitution string. For example, suppose we are trying to extract a value preceded by the string value= from a longer string. Simply substituting the regular expression for the tagged part will retain all the other parts of the string:

```
> str = 'report: 17 value=12 time=2:00'
> sub('value=([^ ]+)','\\1',str)
[1] "report: 17 12 time=2:00"
```

(The regular expression [^]+ is interpreted as one or more occurrences of a character that is not a blank.) By expanding the regular expression to include all the unwanted parts, the substitution will extract just what we want:

```
> sub('^.*value=([^ ]+).*$','\\1',str)
[1] "12"
```

Another strategy is to use regexpr or gregexpr to find the location of the match, and apply sub or gsub to the extracted parts:

```
> str = 'report: 17 value=12 time=2:00'
> greg = gregexpr('value=[^ ]+',str)[[1]]
> sub('value=([^ ]+)','\\1',
+      substring(str,greg,greg
                    + attr(greg, 'match.length') - 1))
[1] "12"
```

8

Data Aggregation

R provides a wide array of functions to aid in aggregating data. For simple tabulation and cross-tabulation, the `table` function is available. For more complex tasks, the available functions can be broken down into two groups: those that are designed to work effectively with arrays and/or lists, like `apply`, `sweep`, `mapply`, `sapply`, and `lapply`, and those that are oriented toward data frames (like `aggregate` and `by`). There is considerable overlap between the two tools, and the output of one can be converted to the equivalent of the output from another, so often the choice of an appropriate function is a matter of personal taste.

We'll start by looking at the `table` function, and then study the other functions which can be used to aggregate data from various sources.

8.1 `table`

The arguments to the `table` function can either be individual vectors representing the levels of interest, or a list or data frame composed of such vectors. The result from `table` will always be an array of as many dimensions as the number of vectors being tabulated, with dimnames extracted from the levels of the cross-tabulated variables. By default, table will not include missing values in its output; to override this, use the `exclude=NULL` argument. When passed a single vector of values, `table` returns an object of class `table`, which can be treated as a named vector. For simple queries regarding individual levels of a tabulated variable, this may be the most convenient form of displaying and storing the values:

```
> pets = c('dog','cat','duck','chicken','duck','cat','dog')
> tt = table(pets)
```

```
> tt
pets
    cat chicken    dog    duck
      2      1      2      2
> tt['duck']
duck
   2
> tt['dog']
dog
  2
```

Alternatively, the output from `table` can be converted to a data frame using `as.data.frame`:

```
> as.data.frame(tt)
     pets Freq
1     cat    2
2 chicken    1
3     dog    2
4    duck    2
```

When multiple vectors are passed to `table`, an array of as many dimensions as there are vectors is returned. For this example, the `state.region` and `state.x77` datasets are used, creating a table that shows the number of states whose income is above and below the median income for all states, broken down by region:

```
> hiinc = state.x77[,'Income'] > median(state.x77[,'Income'])
> stateinc = table(state.region,hiinc)
> stateinc
                 hiinc
state.region    FALSE TRUE
   Northeast        4    5
   South           12    4
   North Central    5    7
   West             4    9
```

This result can be converted to a data frame using `as.data.frame`:

```
> as.data.frame(stateinc)
    state.region hiinc Freq
1      Northeast FALSE    4
2          South FALSE   12
3  North Central FALSE    5
4           West FALSE    4
5      Northeast  TRUE    5
6          South  TRUE    4
7  North Central  TRUE    7
8           West  TRUE    9
```

When passed a data frame, `table` treats each column as a separate variable, resulting in a table that effectively counts how often each row appears in the data frame. This can be especially useful when the result of `table` is passed to `as.data.frame`, since its form will be similar to the input data frame. To illustrate, consider this small example:

```
> x = data.frame(a=c(1,2,2,1,2,2,1),b=c(1,2,2,1,1,2,1),
+                     c=c(1,1,2,1,2,2,1))
> x
  a b c
1 1 1 1
2 2 2 1
3 2 2 2
4 1 1 1
5 2 1 2
6 2 2 2
7 1 1 1
> as.data.frame(table(x))
  a b c Freq
1 1 1 1    3
2 2 1 1    0
3 1 2 1    0
4 2 2 1    1
5 1 1 2    0
6 2 1 2    1
7 1 2 2    0
8 2 2 2    2
```

Since the data frame was formed from a table, all possible combinations, including those with no observations, are included.

Sometimes it is helpful to display the margins of a table, that is, the sum of each row and/or column, in order to understand differences among the levels of the variables from which the table was formed. The `addmargins` function accepts a table and returns a similar table, with the requested margins added. To specify which dimensions should have margins added, the `margin=` argument accepts a vector of dimensions; a value of 1 in this vector means a new row with the margins for the columns will be added, and a value of 2 corresponds to a new column containing row margins. The default operation to create the margins is to use the `sum` function. If some other function is desired, it can be specified through the `FUN=` argument. When a margin is added, the dimnames for the table are adjusted to include a description of the margin. As an example of the use of `addmargins`, consider the `infert` dataset, which contains information about the education and parity of experimental subjects. First, we can generate a cross-tabulation in the usual way:

```
> tt = table(infert$education,infert$parity)
```

```
> tt
```

```
          1  2  3  4  5  6
0-5yrs    3  0  0  3  0  6
6-11yrs  42 42 21 12  3  0
12+ yrs  54 39 15  3  3  2
```

To add a row of margins, we can use the following call to `addmargins`:

```
> tt1 = addmargins(tt,1)
> tt1
```

```
          1  2  3  4  5  6
0-5yrs    3  0  0  3  0  6
6-11yrs  42 42 21 12  3  0
12+ yrs  54 39 15  3  3  2
Sum      99 81 36 18  6  8
```

To add margins to both rows and columns, use a `margin=` argument of `c(1,2)`:

```
> tt12 = addmargins(tt,c(1,2))
> tt12
```

```
           1   2   3   4   5   6 Sum
0-5yrs     3   0   0   3   0   6  12
6-11yrs   42  42  21  12   3   0 120
12+ yrs   54  39  15   3   3   2 116
Sum       99  81  36  18   6   8 248
> dimnames(tt12)
[[1]]
[1] "0-5yrs"  "6-11yrs" "12+ yrs" "Sum"

[[2]]
[1] "1"    "2"    "3"    "4"    "5"    "6"    "Sum"
```

Notice that the dimnames for the table have been updated.

When it's desired to have a table of proportions instead of counts, one strategy would be to use the `sweep` function (Section 8.4) dividing each row and column by its corresponding margin. The `prop.table` function provides a convenient wrapper around this operation. `prop.table` accepts a table, and a `margin=` argument, and returns a table of proportions. With no value specified for `margin=`, the sum of all the cells in the table will be 1; with `margin=1`, each row of the resulting table will add to 1, and with `margin=2`, each column will add to 1. Continuing with the previous example, we can convert our original table to one containing proportions, having each column add to 1, as follows:

```
> prop.table(tt,2)
```

```
                 1       2       3       4       5       6
0-5yrs     0.03030 0.00000 0.00000 0.16667 0.00000 0.75000
6-11yrs    0.42424 0.51852 0.58333 0.66667 0.50000 0.00000
12+ yrs    0.54545 0.48148 0.41667 0.16667 0.50000 0.25000
```

For tables with more than two dimensions, it may be useful to present the table in a "flattened" form using the `ftable` function. To illustrate, consider the `UCBAdmissions` dataset, which is already a table with counts for admission to various departments based on gender. As a three-dimensional table, it would normally be displayed as a series of two-dimensional tables. Using `ftable`, the same information can be displayed in a more compact form:

```
> ftable(UCBAdmissions)
                 Dept   A   B   C   D   E   F
Admit    Gender
Admitted Male          512 353 120 138  53  22
         Female         89  17 202 131  94  24
Rejected Male          313 207 205 279 138 351
         Female         19   8 391 244 299 317
```

The `xtabs` function can produce similar results to the `table` function, but uses the formula language interface. For example, the state income by region table could be reproduced using statements like these:

```
> xtabs(~state.region + hiinc)
                hiinc
state.region    FALSE TRUE
   Northeast        4    5
   South           12    4
   North Central    5    7
   West             4    9
```

If a variable is given on the left-hand side of the tilde (~), it is interpreted as a vector of counts corresponding to the values of the variables on the right-hand side, making it very easy to convert already tabulated data into R's notion of a table:

```
> x = data.frame(a=c(1,2,2,1,2,2,1),b=c(1,2,2,1,1,2,1),
+                c=c(1,1,2,1,2,2,1))
> dfx = as.data.frame(table(x))
> xtabs(Freq ~ a + b + c,data=dfx)
, , c = 1

   b
a   1 2
  1 3 0
  2 0 1
```

```
, , c = 2

    b
a   1 2
    1 0 0
    2 1 2
```

8.2 Road Map for Aggregation

When confronted with an aggregation problem, there are three main considerations:

1. How are the groups that divide the data defined?
2. What is the nature of the data to be operated on?
3. What is the desired end result?

Thinking about these issues will help to point you to the most effective solution for your needs. The following paragraphs should help you make the best choice.

Groups defined as list elements. If the groups you're interested in are already organized as elements of a list, then `sapply` or `lapply` (Section 8.3) are the appropriate functions; they differ in that `lapply` always returns a list, while `sapply` may simplify its output into a vector or array if appropriate. This is a very flexible approach, since the entire data frame for each group is available. Sometimes, if other methods are inappropriate, you can first use the `split` function to create a suitable list for use with `sapply` or `lapply` (Section 8.5).

Groups defined by rows or columns of a matrix. When the goal is to operate on each column or row of a matrix, the `apply` function (Section 8.4) is the logical choice. `apply` will usually return its results as a vector or array, but will return a list if the results of operating on the rows or columns are of different dimensions.

Groups based on one or more grouping variables. A wide array of choices is available for the very common task of operating on subsets of data based on the value of a grouping variable. If the computations you desire each involve only a single vector and produce a single scalar as a result (like calculating a scalar-valued statistic for a variable or set of variables), the `aggregate` function (Section 8.5) is the most likely choice. Since aggregate always returns a data frame, it is especially useful if the desired result is to create a plot or fit a statistical model to the aggregated data.

If your computations involve a single vector, but the result is a vector (for example, a set of quantiles or a vector of different statistics), `tapply` (Section 8.5) is one available option. Unlike `aggregate`, `tapply` returns its results in a vector or array for which individual elements are easy to access, but may produce a difficult-to-interpret display for complex problems. Another approach to the problem is provided by the `reshape` package, available through CRAN, and documented in Section 8.6. It uses a formula interface, and can produce output in a variety of forms.

When the desired result requires access to more than one variable at a time (for example, calculating a correlation matrix, or creating a scatter plot), row indices can be passed to `tapply` to extract the appropriate rows corresponding to each group. Alternatively, the `by` function can be used. Unlike `tapply`, the special list returned by `by` has a print method which will always produce an easily-readable display of the aggregation, but accessing individual elements of the returned list may be inconvenient. Naturally, for tasks like plotting, there is no clear reason to choose one approach over the other.

As mentioned previously, using `split` and `sapply`/`lapply` is a good solution if you find that other methods don't provide the flexibility you need. Finally, if nothing else seems to work, you can write a loop to iterate over the values returned by `unique` or `intersection`, and perform whatever operations you desire. If you take this route, make sure to consider the issues about memory management in loops found in Section 8.7.

8.3 Mapping a Function to a Vector or List

Although most functions in R will automatically operate on each element of a vector, the same is not true for lists. Since many R functions return lists, it's often useful to process each list element in the same way that R naturally does for vectors. To handle situations like this, R provides two functions: `lapply` and `sapply`. Each of these functions takes a list or vector as its first argument, and a function to be applied to each element as its second argument. The difference between the two functions is that `lapply` will always return its result as a list, while `sapply` will simplify its output to a vector or matrix if possible. For example, suppose we have a vector of character strings, and we want to find out how many words are in each vector. Like most functions in R, the `strsplit` function will operate on each element of a vector, returning for each element a new vector containing the individual pieces of that element:

```
> text = c('R is a free environment for statistical analysis',
+             'It compiles and runs on a variety of platforms',
+             'Visit the R home page for more information')
> result = strsplit(text,' ')
> result
[[1]]
[1] "R"           "is"          "a"
[4] "free"        "environment" "for"
[7] "statistical" "analysis"
```

```
[[2]]
[1] "It"        "compiles"   "and"
[4] "runs"      "on"         "a"
[7] "variety"   "of"         "platforms"


[[3]]
[1] "Visit"     "the"        "R"
[4] "home"      "page"       "for"
[7] "more"      "information"
```

Since each vector could potentially contain different numbers of words, strsplit puts its result into a list. The length function will not automatically operate on each list element; instead, it properly reports the number of elements in the returned list:

```
> length(result)
[1] 3
```

To find the length of the individual elements, we can use either sapply or lapply; since the length of each element will be a scalar, sapply would be most appropriate:

```
> nwords = sapply(result,length)
> nwords
[1] 8 9 8
```

Another important use of sapply relates to data frames. When treated as a list, each column of a data frame retains its mode and class. Suppose we're working with the built-in ChickWeight data frame, and we wish to learn more about the nature of each column. Simply using the class function on the data frame will give information about the data frame, not the individual columns:

```
> class(ChickWeight)
[1] "nfnGroupedData"  "nfGroupedData"
[3] "groupedData"     "data.frame"
```

To get the same information for each variable, use sapply:

```
> sapply(ChickWeight,class)
$weight
[1] "numeric"

$Time
[1] "numeric"

$Chick
[1] "ordered" "factor"

$Diet
[1] "factor"
```

Notice that in this case, since the class for `Chick` was of length 2, `sapply` returned its result as a list. This will always be the case when the structure of the data would be lost if `sapply` tried to simplify it into a vector or array.

This same idea can be used to extract columns of a data frame that meet a particular condition. For example, to create a data frame containing only numeric variables, we could use

```
df[,sapply(df,class) == 'numeric']
```

`sapply` or `lapply` can be used as an alternative to loops for performing repetitive tasks. When you use these functions, they take care of the details of deciding on the appropriate form of the output, and eliminate the need to incrementally build up a vector or matrix to store the result. To illustrate, suppose that we wish to generate matrices of random numbers and determine the highest correlation coefficient between any of the variables in the matrix. The first step is to create a function that will generate a single matrix and calculate the maximum correlation coefficient:

```
maxcor = function(i,n=10,m=5){
    mat = matrix(rnorm(n*m),n,m)
    corr = cor(mat)
    diag(corr) = NA
    max(corr,na.rm=TRUE)
}
```

Since `sapply` will always pass an argument to the applied function, a dummy argument (`i`) is added to the function. Since the diagonal of a correlation matrix will always be 1, the diagonal elements of the correlation matrix were masked by assigning them values of `NA`. Suppose we want to generate 1000 100×5 matrices, and find the average value of the maximum correlation:

```
> maxcors = sapply(1:1000,maxcor,n=100)
> mean(maxcors)
[1] 0.1548143
```

Notice that additional arguments to the function being applied (like `n=100` in this case) are passed to the function by including them in the argument list after the function name or definition.

For simpler simulations of this type, the `replicate` function can be used. This function takes as its first argument the number of replications desired, and as its second argument an expression (not a function!) that calculates the desired statistic for the simulation. For example, we can generate a single t-statistic from two groups of normally distributed observations with the following expression:

```
> t.test(rnorm(10),rnorm(10))$statistic
       t
0.2946709
```

Using replicate, we can generate as many of these statistics as we want:

```
> tsim = replicate(10000,t.test(rnorm(10),rnorm(10))$statistic)
> quantile(tsim,c(0.5,0.75,0.9,0.95,0.99))
      50%        75%        90%        95%        99%
0.00882914 0.69811345 1.36578668 1.74995603 2.62827515
```

8.4 Mapping a function to a matrix or array

When your data has the added organization of an array, R provides a convenient way to operate on each dimension of the data through the apply function. This function requires three arguments: the array on which to perform the operation, an index telling apply which dimension to operate on, and the function to use. Like sapply, additional arguments to the function can be placed at the end of the argument list. For matrices, a second argument of 1 means "operate on the rows", and 2 means "operate on the columns".

One common use of apply is in conjunction with functions like scale, which require summary statistics calculated for each column of a matrix. Without additional arguments , the scale function will subtract the mean of each column and divide by the standard deviation, resulting in a matrix of z-scores. To use other statistics, appropriate vectors of values can be calculated using apply and provided to scale using the center= and scale= arguments. For example, by providing a vector of medians for centering, and a vector of mean average deviations for scaling, an alternative standardization to z-scores can be performed. Using the built-in state.x77 dataset, we could perform such a transformation as follows:

```
> sstate = scale(state.x77,center=apply(state.x77,2,median),
+                         scale=apply(state.x77,2,mad))
```

Similar to sapply, apply will try to return its results in a vector or matrix when appropriate, making it useful in cases where several quantities need to be calculated for each row or column of a matrix. Suppose we wish to produce a matrix containing the number of nonmissing observations, the mean and the standard deviation for each column of a matrix. The first step is writing a function which will return what we want for a single column:

```
summfn = function(x)c(n=sum(!is.na(x)),mean=mean(x),sd=sd(x))
```

Now we can apply the function to a data frame with all numeric columns, or a numeric matrix like state.x77:

```
> x = apply(state.x77,2,sumfun)
```

```
> t(x)
            n      mean           sd
Population 50   4246.4200 4.464491e+03
Income     50   4435.8000 6.144699e+02
Illiteracy 50      1.1700 6.095331e-01
Life Exp   50     70.8786 1.342394e+00
Murder     50      7.3780 3.691540e+00
HS Grad    50     53.1080 8.076998e+00
Frost      50    104.4600 5.198085e+01
Area       50  70735.8800 8.532730e+04
```

This example illustrates another advantage of using `apply` instead of a loop, namely, that `apply` will use names that are present in the input matrix or data frame to properly label the result that it returns.

One further use of `apply` is worth mentioning. If a vector needs to be processed in non-overlapping groups, it is sometimes easiest to temporarily treat the vector as a matrix, and use `apply` to operate on the groups. For example, suppose we wish to take the sum of every three adjacent values in a vector. By first forming a three-column matrix, we can process the groups conveniently using `apply`:

```
> x = 1:12
> apply(matrix(x,ncol=3,byrow=TRUE),1,sum)
[1]   6 15 24 33
```

The `apply` function is very general, and for certain applications, there may be more efficient methods available to perform the necessary computations. For example, if the statistic to be calculated is the sum or the mean, matrix computations will be more efficient than calling `apply` with the appropriate function. In cases like this, the `rowSums`, `colSums`, `rowMeans`, or functions can be used. Each of these functions accepts a matrix (or a data frame which will be coerced to a matrix), and an optional `na.rm=` argument to specify the handling of missing values. Since these functions will accept logical values as input as well as numeric values, they can be very useful for counting operations.

For example, consider the dataset `USJudgeRatings`, which has ratings for 43 judges in twelve categories. To get the mean rating for each category, the `colMeans` function could be used as follows:

```
> mns = colMeans(USJudgeRatings)
> mns
    CONT     INTG     DMNR     DILG     CFMG
7.437209 8.020930 7.516279 7.693023 7.479070
    DECI     PREP     FAMI     ORAL     WRIT
7.565116 7.467442 7.488372 7.293023 7.383721
    PHYS     RTEN
7.934884 7.602326
```

To count the number of categories for which each judge received a score of 8 or greater, the `rowSums` function can be used by providing the appropriate logical matrix:

```
> jscore = rowSums(USJudgeRatings >= 8)
> head(jscore)
 AARONSON,L.H.  ALEXANDER,J.M.  ARMENTANO,A.J.
            1               8               1
    BERDON,R.I.    BRACKEN,J.J.       BURNS,E.B.
           11               0              10
```

A common situation when processing a matrix by rows or columns is that each row or column needs to be processed differently, based on the values of an auxiliary vector which already exists. In cases like this, the `sweep` function can be used. Like `apply`, the first two arguments to `sweep` are the matrix to be operated on and the index of the dimension to be used for repetitive processing. In addition, `sweep` takes a third argument representing the vector to be used when processing each column, and finally a fourth argument providing the function to be used. `sweep` operates by building matrices which can be operated on in a single call, so, unlike `apply`, only functions which can operate on arrays of values can be passed to `sweep`. All of the built-in binary operators, such as addition (`"+"`), subtraction (`"-"`), multiplication (`"*"`), and division (`"/"`) can be used, but, in general, it will be necessary to make sure an arbitrary function will work properly with `sweep`. For example, suppose we have a vector representing the maximum value found in each column of a matrix, and we wish to divide each column of the matrix by its corresponding maximum. Using the `state.x77` data frame, we could use `sweep` as follows:

```
> maxes = apply(state.x77,2,max)
> swept = sweep(state.x77,2,maxes,"/")
> head(swept)
            Population     Income Illiteracy  Life Exp     Murder
Alabama     0.17053496 0.5738717  0.7500000 0.9381793 1.0000000
Alaska      0.01721861 1.0000000  0.5357143 0.9417120 0.7483444
Arizona     0.10434947 0.7173397  0.6428571 0.9585598 0.5165563
Arkansas    0.09953769 0.5349169  0.6785714 0.9600543 0.6688742
California  1.00000000 0.8098179  0.3928571 0.9743207 0.6821192
Colorado    0.11986980 0.7733967  0.2500000 0.9790761 0.4503311
              HS Grad      Frost       Area
Alabama     0.6136701 0.10638298 0.08952178
Alaska      0.9910847 0.80851064 1.00000000
Arizona     0.8632987 0.07978723 0.20023057
Arkansas    0.5928678 0.34574468 0.09170562
California  0.9301634 0.10638298 0.27604549
Colorado    0.9494799 0.88297872 0.18319233
```

Now suppose that we wish to calculate the mean value for each variable using only those values which are greater than the median for that variable. We can calculate the medians using `apply` and then write a simple function to find the mean of the values we're interested in.

```
> meds = apply(state.x77,2,median)
> meanmed = function(var,med)mean(var[var>med])
> meanmed(state.x77[,1],meds[1])
[1] 7136.16
> meanmed(state.x77[,2],meds[2])
[1] 4917.92
```

Although the function works properly for individual columns, it returns only a single value when used in conjunction with `sweep`:

```
> sweep(state.x77,2,meds,meanmed)
[1] 15569.75
```

The source of the problem is the inequality used to subset the variable values, since it will not properly operate on the array that `sweep` produces to calculate its results. In cases like this, the `mapply` function can be used. By converting the input matrix to a data frame, each variable in the input will be processed in parallel to the vector of medians, providing the desired result:

```
> mapply(meanmed,as.data.frame(state.x77),meds)
[1]    7136.160    4917.920      1.660     71.950     10.544
[6]      59.524    146.840  112213.400
```

By default, `mapply` will always simplify its results, as in the previous case where it consolidated the results in a vector. To override this behavior, and return a list with the results of applying the supplied function, use the `SIMPLIFY=FALSE` argument.

8.5 Mapping a Function Based on Groups

To calculate scalar data summaries of one or more columns of a data frame or matrix, the `aggregate` function can be used. Although this function is limited to returning scalar values, it can operate on multiple columns of its input argument, making it a natural choice for data summaries for multiple variables. The first argument to `aggregate` is a data frame or matrix containing the variables to be summarized, the second argument is a list containing the variables to be used for grouping, and the third argument is the function to be used to summarize the data. For example, the `iris` dataset contains the values of four variables measured on a variety of samples from three species of irises. To find the means of all four variables broken down by species, `aggregate` can be called as follows:

```
> aggregate(iris[-5],iris[5],mean)
      Species Sepal.Length Sepal.Width Petal.Length Petal.Width
1      setosa        5.006       3.428        1.462       0.246
2 versicolor        5.936       2.770        4.260       1.326
3  virginica        6.588       2.974        5.552       2.026
```

Since the second argument must be a list, when a data frame is being processed it is often convenient to refer to the grouping columns using single bracket subscripts, since columns accessed this way will naturally be in the form of a list. In addition, with more than one grouping variable, specifying the columns this way will insure that the grouping variables' names will be automatically transfered to the output data frame. If the columns are passed as manually constructed list, aggregate will use names like Group.1 to identify the grouping variables, unless names are provided for the list elements.

As an example, suppose we wish to calculate the mean weight for observations in the ChickWeight data frame, broken down by the variables Time and Diet. Specifying the grouping variables as ChickWeight[c('Time','Diet')] will result in the grouping columns being properly labeled:

```
> cweights = > aggregate(ChickWeight$weight,
+                          ChickWeight[c('Time','Diet')],mean)
> head(cweights)
  Time Diet        x
1    0    1 41.40000
2    2    1 47.25000
3    4    1 56.47368
4    6    1 66.78947
5    8    1 79.68421
6   10    1 93.05263
```

Alternatively, a constructed list like

```
list(Time=ChickWeight$Time,Diet=ChickWeight$Diet)
```

could be used to achieve the same result.

To process a single vector based on the values of one or more grouping vectors, the tapply function can also be used. The returned value from tapply will be an array with as many dimensions as there were vectors that defined the groups. For example, the PlantGrowth dataset contains information about the weight of plants receiving one of three different treatments. To find the maximum weight for plants exposed to each of the treatments, we could use tapply as follows:

```
> maxweight = tapply(PlantGrowth$weight,PlantGrowth$group,max)
> maxweight
ctrl trt1 trt2
6.11 6.03 6.31
```

Since there was only one grouping factor, the results were returned in the form of a named vector. To convert this vector into a data frame, it can temporarily be converted into a table using `as.table`, and then passed to `as.data.frame`, since there is a special method for converting tables into data frames:

```
> as.data.frame(as.table(maxweight))
  Var1 Freq
1 ctrl 6.11
2 trt1 6.03
3 trt2 6.31
```

To use a name other than `Freq` in the data frame, `as.data.frame.table` can be called directly, using the `responseName=` argument:

```
> as.data.frame.table(as.table(maxweight),
                       responseName='MaxWeight')
  Var1 MaxWeight
1 ctrl      6.11
2 trt1      6.03
3 trt2      6.31
```

Unlike `aggregate`, `tapply` is not limited to returning scalars. For example, if we wanted the range of weights for each group in the `PlantGrowth` dataset, we could use

```
> ranges = tapply(PlantGrowth$weight,PlantGrowth$group,range)
> ranges
$ctrl
[1] 4.17 6.11

$trt1
[1] 3.59 6.03

$trt2
[1] 4.92 6.31
```

In this case. `tapply` returns a named array of vectors. Individual elements can be accessed in the usual way:

```
> ranges[[1]]
[1] 4.17 6.11
> ranges[['trt1']]
[1] 3.59 6.03
```

To convert values like this to data frames, the `dimnames` of the returned object can be combined with the values. When each element of the vector is of the same length, this operation is fairly straightforward, but the problem becomes difficult when the return values are of different lengths. In the current example, we can convert the values to a numeric matrix, and then form a data frame by combining the matrix with the `dimnames`:

```
> data.frame(group=dimnames(ranges)[[1]],
+              matrix(unlist(ranges),ncol=2,byrow=TRUE))
  group   X1   X2
1  ctrl 4.17 6.11
2  trt1 3.59 6.03
3  trt2 4.92 6.31
```

data.frame was used here instead of cbind to prevent the numeric values from being coerced to character values when they were combined with the levels of the grouping variable.

When more than one grouping variable is used with tapply, and the return value from the function used is not a scalar, the returned object is somewhat more difficult to interpret. For example, the CO2 dataset contains information about the uptake of carbon dioxide by different types of plants exposed to different treatments. Suppose we were interested in the range of CO_2 uptake for plants of each type and treatment. We can call tapply as follows:

```
> ranges1 = tapply(CO2$uptake,CO2[c('Type','Treatment')],range)
> ranges1
             Treatment
Type          nonchilled chilled
  Quebec       Numeric,2 Numeric,2
  Mississippi Numeric,2 Numeric,2
```

The returned value is a matrix of lists, which explains the unusual form of the output when we display the object. Individual elements can still be accessed as expected:

```
> ranges[['Quebec','chilled']]
[1]   9.3 42.4
```

Such objects can be converted to data frames by applying expand.grid (see Section 2.8.1) to the dimnames before combining them with the values:

```
> data.frame(expand.grid(dimnames(ranges1)),
+              matrix(unlist(ranges1),byrow=TRUE,ncol=2))
          Type Treatment   X1   X2
1       Quebec nonchilled 13.6 45.5
2  Mississippi nonchilled 10.6 35.5
3       Quebec    chilled  9.3 42.4
4  Mississippi    chilled  7.7 22.2
```

The function argument to tapply is not required; calling tapply without a function will return a vector of indices which can be used as a subscript to the array of values that tapply produces when a function is provided. For example, suppose we wish to subtract the median value of the uptake variable in the CO2 data frame, where the median is calculated separately for each Type/Treatment combination. The first step is calculating the medians for each group using tapply:

```
> meds = tapply(CO2$uptake,CO2[c('Type','Treatment')],median)
```

Next, the indices are calculated using an identical call to `tapply` without a function, and they are used as a subscript to the median vector:

```
> inds = tapply(CO2$uptake,CO2[c('Type','Treatment')])
> inds
 [1] 1 1 1 1 1 1 1 1 1 1 1 1 1 1 1 1 1 1 1 1 1 1 1 3 3 3 3 3 3 3 3
[31] 3 3 3 3 3 3 3 3 3 3 3 3 2 2 2 2 2 2 2 2 2 2 2 2 2 2 2 2 2 2
[61] 2 2 2 4 4 4 4 4 4 4 4 4 4 4 4 4 4 4 4 4 4 4 4 4
> adj.uptake = CO2$uptake - meds[inds]
```

The `ave` function combines these two operations in a single function call:

```
> adj.uptake = CO2$uptake -
+         ave(CO2$uptake,CO2[c('Type','Treatment')],FUN=median)
```

Since `ave` can accept multiple grouping variables, the function to be used for summarization must be identified using `FUN=`. Thus, the previous example could have been carried out with the following statement:

```
> adj.uptake = CO2$uptake -
+         ave(CO2$uptake,CO2$Type,CO2$Treatment,FUN=median)
```

When more than a single vector needs to be processed, a variety of options is available. To put the problem into context, consider the task of finding the maximum eigenvalue of the correlation matrices of the four variables from the `iris` dataset, broken down by the species of the plant. One solution is to use the `split` function, which takes a data frame and a list of grouping variables and returns a list containing data frames representing the observations for each level of the grouping variables. Such a list can then be processed using `sapply` or `lapply` to provide the final result. When working with problems like this, the first step is usually defining a function to provide the required result for a single data frame. In this case, an appropriate function could be written as follows:

```
> maxeig = function(df)eigen(cor(df))$val[1]
```

Next, the numeric values in the data frame can be passed to `split` to provide a list of data frames for further processing:

```
> frames = split(iris[-5],iris[5])
```

Finally, this result can be passed to `sapply` along with the function to do the work:

```
> sapply(frames,maxeig)
   setosa versicolor  virginica
 2.058540   2.926341   2.454737
```

As always, these operations can be condensed to a single expression, although there is no great advantage in doing so.

```
> sapply(split(iris[-5],iris[5]),
+          function(df)eigen(cor(df))$val[1])
   setosa versicolor  virginica
  2.058540   2.926341   2.454737
```

A less direct, but sometimes useful solution involves passing a vector of row indices to `tapply` and modifying the function used to calculate the maximum eigenvalue to operate on selected rows of the data:

```
> tapply(1:nrow(iris),iris['Species'],
+          function(ind,data)eigen(cor(data[ind,-5]))$val[1],
          data=iris)
Species
   setosa versicolor  virginica
  2.058540   2.926341   2.454737
```

Finally, the `by` function can be used. This generalizes the idea of `tapply` to operate on entire data frames broken down by a list of grouping variables. Thus, the first argument to `by` is a data frame, and the remaining arguments are similar to those of `tapply`. For the eigenvalue problem, a solution using `by` is as follows:

```
> max.e = by(iris,iris$Species,
+                    function(df)eigen(cor(df[-5]))$val[1])
> max.e
iris$Species: setosa
[1] 2.058540
------------------------------------------------------------
iris$Species: versicolor
[1] 2.926341
------------------------------------------------------------
iris$Species: virginica
[1] 2.454737
```

In this case, `by` returned a scalar, so the result can be converted to a data frame by using a combination of `as.table` and `as.data.frame`:

```
> as.data.frame(as.table(max.e))
   iris.Species      Freq
1       setosa 2.058540
2   versicolor 2.926341
3    virginica 2.454737
```

When there are multiple variables describing the groups to be processed, the result from `by` needs additional processing to get it in the form of a data frame. Consider again the CO_2 dataset. Suppose we wish to find the number of observations, mean, and standard deviation of the variable `uptake`, broken down by `Type` and `Treatment` combinations. First, a simple function to return the required values is written. By putting together the values with `data.frame` instead of c, we insure that the mode of the numeric results will be preserved after we combine them with the level information for the grouping variables:

```
> sumfun = function(x)data.frame(n=length(x$uptake),
+                         mean=mean(x$uptake),sd=sd(x$uptake))
> bb = by(CO2,CO2[c('Type','Treatment')],sumfun)
> bb
Type: Quebec
Treatment: nonchilled
   n    mean       sd
1 21 35.33333 9.59637
-----------------------------------------------------------------
Type: Mississippi
Treatment: nonchilled
   n    mean       sd
1 21 25.95238 7.402136
-----------------------------------------------------------------
Type: Quebec
Treatment: chilled
   n    mean       sd
1 21 31.75238 9.644823
-----------------------------------------------------------------
Type: Mississippi
Treatment: chilled
   n    mean       sd
1 21 15.81429 4.058976
```

Each of the rows returned by the by function is in the form that we would like for a data frame containing these results, so it would be natural to use rbind to convert this result to a data frame; however, it is tedious to pass each row to the rbind function individually. In cases like this, the do.call function, first introduced in Section 6.5, can usually generalize the operation so that it will be carried out properly regardless of how many elements need to be processed. Recall that do.call takes a list of arguments and passes them to a function as if they were the argument list for the function call. In this example, the call to do.call is as follows:

```
> do.call(rbind,bb)
    n    mean       sd
1  21 35.33333 9.596371
11 21 25.95238 7.402136
12 21 31.75238 9.644823
13 21 15.81429 4.058976
```

With two grouping variables, the names and levels of the grouping factors are not present in the result. This can be remedied by combining a call to `expand.grid` with the previous result. Since all the parts being combined are data frames, they can be safely combined using `cbind`:

```
> cbind(expand.grid(dimnames(bb)),do.call(rbind,bb))
          Type   Treatment  n     mean       sd
1        Quebec nonchilled 21 35.33333 9.596371
2 Mississippi nonchilled 21 25.95238 7.402136
3        Quebec    chilled 21 31.75238 9.644823
4 Mississippi    chilled 21 15.81429 4.058976
```

8.6 The reshape Package

An alternative approach to aggregation is provided by the **reshape** package, available from CRAN. The functions in this package provide a unified approach to aggregation, based on an extended formula notation. The core idea behind the **reshape** package is to create a "melted" version of a dataset (through the `melt` function), which can then be "cast" (with the `cast` function) into an object with the desired orientation. To melt a data frame, list, or array into the appropriate melted form, it is first necessary to divide the variables into id variables and measure or analysis variables; this should generally be obvious from the nature of the data. By default, `melt` treats factor and integer variables as id variables, and the remaining variables as analysis variables; if your data is structured according to this convention, no additional information needs to be provided to `melt`. Otherwise, the `id.var=` or `measure.var=` arguments can be used; if you specify one, it will assume all the other variables are of the other type. Once a dataset is melted, it can be cast into a variety of forms.

As a simple example, consider a dataset formed from the `state.x77` data frame, combined with the `state.region` variable:

```
> states = data.frame(state.x77,state=row.names(state.x77),
+                     region=state.region,row.names=1:50)
```

The `state` and `region` variables are stored as factors, so they will be automatically recognized as id variables when we melt the data:

```
> library(reshape)
> mstates = melt(states)
Using state, region as id variables
```

Notice that `melt` displays the names of variables that have been automatically assigned as id variables. The basic `melting` operation preserves the id variables, and converts the measured variables into two columns named `variable` (which identifies which variable is being measured) and `value` (which contains the actual values). You can use a name other than `variable` by specifying a `variable_name=` argument to `melt`.

The left-hand side of the formula passed to `cast` represents the variable(s) which will appear in the columns of the result, and the right-hand side describes the variables which will appear in the rows. Formulas used by `cast` can include a single dot (.) to represent an overall summary, or three dots ... to represent all variables not otherwise included in the formula. In the simplest case, we can reproduce the original dataset with a formula like "... ~ variable".

When used for aggregation, an aggregation function should be supplied; if not it defers to using `length`. Suppose we wish to find the mean for each variable, broken down by region, with the regions appearing as a column in the output data frame:

```
> cast(mstates,region~variable,mean)
         region Population   Income Illiteracy Life.Exp
1     Northeast   5495.111 4570.222   1.000000 71.26444
2         South   4208.125 4011.938   1.737500 69.70625
3 North Central   4803.000 4611.083   0.700000 71.76667
4          West   2915.308 4702.615   1.023077 71.23462
      Murder  HS.Grad    Frost      Area
1   4.722222 53.96667 132.7778  18141.00
2  10.581250 44.34375  64.6250  54605.12
3   5.275000 54.51667 138.8333  62652.00
4   7.215385 62.00000 102.1538 134463.00
```

If we wanted a separate row for each `variable` instead of each `region`, we can reverse the role of those variables in the formula:

```
> cast(mstates,variable~region,mean)
       variable    Northeast       South North Central
1    Population  5495.111111  4208.12500    4803.00000
2        Income  4570.222222  4011.93750    4611.08333
3    Illiteracy     1.000000     1.73750       0.70000
4      Life.Exp    71.264444    69.70625      71.76667
5        Murder     4.722222    10.58125       5.27500
6       HS.Grad    53.966667    44.34375      54.51667
7         Frost   132.777778    64.62500     138.83333
8          Area 18141.000000 54605.12500   62652.00000
           West
1 2.915308e+03
2 4.702615e+03
3 1.023077e+00
4 7.123462e+01
5 7.215385e+00
6 6.200000e+01
7 1.021538e+02
8 1.344630e+05
```

To limit the variables that are used, we can use the `subset=` argument of `cast`. Since this argument uses the melted data, we need to refer to the variable named `variable`:

```
> cast(mstates,region~variable,mean,
+             subset=variable %in% c('Population','Life.Exp'))
        region Population Life.Exp
1      Northeast   5495.111 71.26444
2          South   4208.125 69.70625
3 North Central   4803.000 71.76667
4           West   2915.308 71.23462
```

Unlike the `aggregate` function which does not accept functions which return vectors of values, `cast` allows such functions, and uses the names of the returned vector to form new variable names in its output. Alternatively, a list of functions can be provided. Suppose we wish to calculate the mean, median, and standard deviations for `Population` and `Lif.Exp` in the `states` data frame. Since built-in functions exist for each statistic, they can be passed to `cast` as a list: First, we can calculate these quantities for the entire dataset:

```
> cast(mstates,.~variable,c(mean,median,sd),
+             subset=variable %in% c('Population','Life.Exp'))
  value Population_mean Population_median Population_sd
1 (all)          4246.42            2838.5       4464.491
    Life.Exp_mean Life.Exp_median Life.Exp_sd
1        70.8786          70.675     1.342394
```

Since `variable` was specified on the right-hand side of the tilde, all of the statistics for all of the variables are listed in a single row. A more familiar form would have the variables listed in a column, once again achieved by reversing the roles of the variables in the formula:

```
> cast(mstates,variable~.,c(mean,median,sd),
+       subset=variable %in% c('Population','Life.Exp'))
    variable       mean   median         sd
1 Population 4246.4200 2838.500 4464.491433
2   Life.Exp   70.8786   70.675    1.342394
```

To aggregate using a grouping variable, the period in the formula can be replaced by the grouping variable, in this case `region`:

```
> cast(mstates,region~variable,c(mean,median,sd),
+             subset=variable %in% c('Population','Life.Exp'))
        region Population_mean Population_median Population_sd
1      Northeast         5495.111            3100.0      6079.565
2          South         4208.125            3710.5      2779.508
3 North Central         4803.000            4255.0      3702.828
4           West         2915.308            1144.0      5578.607
    Life.Exp_mean Life.Exp_median Life.Exp_sd
```

1	71.26444	71.23	0.7438769
2	69.70625	70.07	1.0221994
3	71.76667	72.28	1.0367285
4	71.23462	71.71	1.3519715

If the roles of `region` and `variable` were reversed, there would be one variable for each combination of `region` and `mean`, `median`, and `sd`, which might not be convenient for display or further manipulation. To provide added flexibility, the vertical bar (|) can be used to cause `cast` to produce a list instead of a data frame. To create a list with a separate data summary for each region, we can specify `region` after the vertical bar, and replace it with a period in the formula:

```
> cast(mstates,variable~.|region,
+               c(mean,median,sd),
+               subset=variable%in%c('Population','Life.Exp'))
$Northeast
       variable       mean  median          sd
1 Population 5495.11111 3100.00 6079.5651457
2    Life.Exp   71.26444   71.23   0.7438769

$South
       variable       mean  median          sd
1 Population 4208.12500 3710.50 2779.508251
2    Life.Exp   69.70625   70.07   1.022199

$'North Central'
       variable       mean  median          sd
1 Population 4803.00000 4255.00 3702.827593
2    Life.Exp   71.76667   72.28   1.036729

$West
       variable       mean  median          sd
1 Population 2915.30769 1144.00 5578.607015
2    Life.Exp   71.23462   71.71   1.351971
```

Note that this creates a separate list element for each region, and that the contents of these elements are similar to those created with the formula "variable ~ ." in a previous example.

The principles in the previous example extend readily to the case with more than one id variable. Consider once again the `ChickWeight` data frame. The variables in this dataset are `weight`, `Time`, `Chick`, and `Diet`. The last three variables represent id variables, with `weight` being the only measure variable. Since `Time` is stored as a numeric variable, it is necessary to explicitly provide either the id or measure variables to the `melt` function:

```
> mChick = melt(ChickWeight,measure.var='weight')
```

To create a data frame with the median value of weight for each level of Diet and Time, the following call to cast can be used:

```
> head(cast(mChick,Diet + Time ~ variable,median))
  Diet Time weight
1    1    0     41
2    1    2     49
3    1    4     56
4    1    6     67
5    1    8     79
6    1   10     93
```

Notice that the variable specified last on the left-hand side (Time) is the one that varies the fastest.

To create a separate column for the median at each time, Time can be moved to the right-hand side of the formula:

```
> cast(mChick,Diet ~ Time + variable,mean)
  Diet 0_weight 2_weight 4_weight 6_weight  8_weight
1    1     41.4    47.25 56.47368 66.78947  79.68421
2    2     40.7    49.40 59.80000 75.40000  91.70000
3    3     40.8    50.40 62.20000 77.90000  98.40000
4    4     41.0    51.80 64.50000 83.90000 105.60000
  10_weight 12_weight 14_weight 16_weight 18_weight 20_weight
1  93.05263  108.5263  123.3889  144.6471  158.9412  170.4118
2 108.50000  131.3000  141.9000  164.7000  187.7000  205.6000
3 117.10000  144.4000  164.5000  197.4000  233.1000  258.9000
4 126.00000  151.4000  161.8000  182.0000  202.9000  233.8889
  21_weight
1  177.7500
2  214.7000
3  270.3000
4  238.5556
```

To create a list, with one element for each Diet, and the median of weight for each Time, use the vertical bar as follows:

```
> cast(mChick,Time ~ variable|Diet,mean)
$'1'
  Time    weight
1    0  41.40000
2    2  47.25000
3    4  56.47368
4    6  66.78947
5    8  79.68421
6   10  93.05263
```

. . .

```
$'4'
   Time  weight
1     0  41.0000
2     2  51.8000
3     4  64.5000
4     6  83.9000
5     8 105.6000
6    10 126.0000
```

. . .

In the previous example there were valid values for each combination of the id variables. If this is not the case, the default behavior of `cast` is to only include combinations actually encountered in the data. To include all possible combinations, use the `add.missing=TRUE` argument. For example, suppose we remove one combination of `Diet` and `Time` from `ChickWeight`:

```
> xChickWeight = subset(ChickWeight,
+                  !(Diet == 1 & Time == 4))
> mxChick = melt(xChickWeight,measure.var='weight')
> head(cast(mxChick,Diet + Time ~ variable,median))
  Diet Time weight
1    1    0     41
2    1    2     49
3    1    6     67
4    1    8     79
5    1   10     93
6    1   12    106
```

By using `add.missing=TRUE`, observations for the missing combinations will be created, with a missing value for the analysis variable:

```
> head(cast(mxChick,Diet + Time ~ variable,median,
+              add.missing=TRUE))
  Diet Time weight
1    1    0     41
2    1    2     49
3    1    4     NA
4    1    6     67
5    1    8     79
6    1   10     93
```

In each of the preceding examples, the dataset was first `melted`, then repeated calls to `cast` were carried out. If only a single call to `cast` is needed, the `recast` function combines the melt and cast steps into a single call:

```
> head(recast(xChickWeight,measure.var='weight',
```

```
+                 Diet + Time ~ variable,median,
+                 add.missing=TRUE))
  Diet Time weight
1   1    0    41
2   1    2    49
3   1    4    NA
4   1    6    67
5   1    8    79
6   1   10    93
```

8.7 Loops in R

In previous sections, the apply family of functions (and associated wrappers) has been presented as the first choice for most repetitive tasks, such as operating on each element of a list, or performing a computation for nonoverlapping subgroups of the data. The major factor in this decision has to do with the simplicity of the functions, as well as their ability to properly use any names which have been assigned to their input arguments. But this way of programming may be awkward and unfamiliar, and many people would like to leverage their knowledge of other programming languages into R by using more familiar programming constructs like loops. An examination of some of the apply-style functions' source code will show that these functions internally use loops to actually get their work done, so arguments against loops based solely on efficiency do not carry much weight. The real problem with loops is that there are some very intuitive operations that may be implemented with loops that turn out to be extremely inefficient in R. In this and the following sections, we'll access the efficiency of different approaches to common problems with the use of the `system.time` function. This function accepts any valid R expression, and returns a vector of length five, containing the user CPU time, the system CPU time, the elapsed time, and the user and system times from any subprocesses. The first value shown, user CPU, is usually the most useful measure of efficiency, and will vary less than the other values when the same task is repeated several times. Since the argument handling in functions uses equal signs to identify keywords, the one restriction when using `system.time` is that assignment statements which are to be timed must use the "gets" form of the assignment operator, namely, `<-` instead of the equal sign.

Before looking at the cases to avoid, let's consider a simple example: finding the mean of each column of a matrix. This problem is so common that the `rowMeans` function is provided for an extremely efficient solution:

```
> dat = matrix(rnorm(1000000),10000,100)
> system.time(mns <- rowMeans(dat))
[1] 0.008 0.000 0.010 0.000 0.000
```

Another solution is to use `apply`:

```
> system.time(mns <- apply(dat,2,mean))
[1] 0.032 0.020 0.056 0.000 0.000
```

Next, we can use a loop to calculate the mean of each column separately. Notice that in this case, we need to initialize the result vector `mns` to accommodate the answer:

```
> system.time({m <- ncol(dat)
+                for(i in 1:m)mns[i] <- mean(dat[,i])})
[1] 0.032 0.004 0.036 0.000 0.000
```

There really isn't that much of a difference in execution time (the loop uses slightly less system time). The main advantage of `apply` in this case is that it eliminates the need to worry about the result vector, and, if the matrix were named, those names would be passed on to the result.

Keep in mind that the previous example still took advantage of vectorization: each column mean was calculated from a single call to `mean`. It is almost always a mistake to loop over each element of a matrix. Consider the following function, which calculates the mean of each column of the matrix by adding together every element and then dividing by the column length:

```
> slowmean = function(dat){
+    n = dim(dat)[1]
+    m = dim(dat)[2]
+    mns = numeric(m)
+    for(i in 1:n){
+        sum = 0;
+        for(j in 1:m)sum = sum + dat[j]
+        mns[i] = sum / n
+    }
+    return(mns)
+}
> system.time(mns <- slowmean(dat))
[1] 2.100 0.000 2.097 0.000 0.000
```

Without any vectorization, the computation is much slower than the other solutions. This illustrates that unless some kind of vectorization is used, computations in R will be very slow.

Before leaving this problem, it should be mentioned that, for any given problem, there may be unique solutions available. For example, the mean of each column of a matrix can be calculated directly using matrix expressions as follows:

```
> system.time({m = dim(dat)[1];mns = rep(1,m) %*% dat / m})
[1] 0.020 0.000 0.021 0.000 0.000
```

This represents an improvement over the apply and loop-based solutions, but is still not as efficient as the `colMeans` solution.

This illustrates that loops, in and of themselves, are not necessarily inefficient in R, but they should certainly take advantage of any vectorization possible to keep them competitive with other techniques.

To understand the kinds of loops which cause problems in R, it's worthwhile to recall how matrices are stored in R, namely, as a one-dimensional vector, with the columns of the matrix "stacked" on top of each other. A very common operation is to build up a matrix iteratively, by starting with an empty matrix, and using the `rbind` function to grow the matrix one row at a time. There are two problems with this approach. First, the size of the matrix changes at each iteration, requiring additional time to be spent in memory allocations. More importantly, since adding a row changes the size of each column in the matrix, all of the matrix elements need to be rearranged in memory each time a new row is added. These repeated memory allocations and rearrangements very quickly take their toll on the efficiency of a program.

Consider the trivial task of creating a matrix, each of whose rows represent the numbers from 1 to 100. Because of recycling rules, this can be achieved as follows:

```
> system.time(m <- matrix(1:100,10000,100,byrow=TRUE))
[1] 0.022 0.003 0.025 0.000 0.000
```

Performing the same operation by incrementally building the matrix is much slower:

```
> buildrow = function(){
+     res = NULL
+     for(i in 1:10000)res = rbind(res,1:100)
+     res
+ }
> system.time(buildrow())
[1] 239.236  21.446 260.707   0.000   0.000
```

Two forces are slowing the computation: first, the size of `res` is changing each time a new row is added to the matrix, causing R to reallocate memory at each iteration. In addition, since R stores its matrices internally by columns, the addition of a row to the matrix means that every column in the matrix needs to be extended, resulting in large amounts of data being moved around in memory. By this reasoning, it would be faster to build the matrix by columns of equal size, since less rearrangement of the data will be necessary:

```
> buildcol = function(){
+     res = NULL
+     for(i in 1:10000)res = cbind(res,1:100)
+     t(res)
+ }
> system.time(buildcol())
[1] 142.666  20.596 163.289   0.000   0.000
```

While this does represent a speedup, it is still far from an optimal solution. What makes the first technique so fast is that when the `matrix` function is used, the size of the result can be determined before the data is generated. We can provide the same advantage to a loop-based solution as follows:

```
> buildrow1 = function(){
+     res = matrix(0,10000,100)
+     for(i in 1:10000)res[i,] = 1:100
+     res
+ }
> system.time(buildrow1())
[1] 0.242 0.015 0.257 0.000 0.000
```

Even if we didn't know how many rows the matrix would contain, it would still be faster to allocate more rows than we need, and then truncate the matrix at the end. For example, let's include only 50% of the rows by checking the value of a random number before adding that row to the output matrix. First, we'll start with a NULL matrix:

```
> somerow1 = function(){
+     res = NULL
+     for(i in 1:10000)if(runif(1) < .5)res = rbind(res,1:100)
+     res
+ }
> system.time(somerow1())
[1] 51.007  6.062 57.125  0.000  0.000
```

Next, we'll allocate a matrix large enough to hold all the rows, then truncate it at the end:

```
> somerow2 = function(){
+     res = matrix(0,10000,100)
+     k = 0
+     for(i in 1:10000)if(runif(1) < .5){
+         k = k + 1
+         res[k,] = 1:100
+     }
+     res[1:k,]
+ }
> system.time(somerow2())
[1] 0.376 0.027 0.404 0.000 0.000
```

Provided there is enough memory for the initial allocation, creating a sufficiently large matrix before beginning to build it will generally be much faster than repeatedly calling `rbind`.

If a situation arises where it is difficult or impossible to allocate an appropriate matrix before building the rows, we can take advantage of the fact that lists in R are stored very differently than matrices. In particular, the memory used by list elements does not have to be contiguous, which means that adding elements to a list doesn't require as much manipulation of data within memory as the corresponding operation on a matrix. The strategy is to build a list of the rows that will eventually become the matrix, and then use do.call to pass all of the rows to rbind in a single operation:

```
> somerow3 = function(){
+    res = list()
+    for(i in 1:10000)if(runif(1) < .5)res = c(res,list(1:100))
+    do.call(rbind,res)
+ }
> system.time(somerow3())
[1] 33.308  0.247 33.575  0.000   0.000
```

While nowhere near as fast as more optimal methods, this technique may prove useful in those situations where the size of the final result may be difficult to determine.

9

Reshaping Data

R is designed so that individual functions don't have complete flexibility with regard to their inputs. Most functions expect their input data to be arranged in a particular way, and it's the responsibility of the user of the function to make sure that the input data is in an appropriate form. So even after you've read in or created your data, it may be necessary to modify your data to suit a function you need.

The focus of this chapter will be on working with data frames, since that is the form required for the majority of the functions in R.

9.1 Modifying Data Frame Variables

Since data frames are lists, new variables can be created by simply assigning their value to a column that doesn't already exist in the data frame. Since operations in R are vectorized, transformations can be carried out without the need to use loops. For example, consider the `Loblolly` data frame which has variables for `height` and `age` for a number of trees. To create a variable called `logheight` representing the log of the `height` variable, we could use statements like

```
> Loblolly$logheight = log(Loblolly$height)
```

or

```
> Loblolly['logheight'] = log(Loblolly['height'])
```

The system's version of the `Loblolly` data frame will not be changed by these statements, but your local copy of `Loblolly` will have the new `logheight` column.

Two functions are handy to avoid the need of retyping the data frame name in order to access columns of a data frame. The `with` function can be used to evaluate any expression, first looking in a data frame of your choice to resolve variables. For example, the `logheight` column in the previous example could be created using

```
> with(Loblolly,log(height))
```

In cases where new columns are being added to an existing data frame, the `transform` function can be used. The first argument to `transform` is a data frame, and the remaining arguments define new columns which will be returned along with all the columns of the original data frame. Each new column is defined by a `name=value` pair. So an alternate way of creating the `logheight` column in the Loblolly dataset would be

```
> Loblolly = transform(Loblolly,logheight = log(height))
```

Once again, the system version of `Loblolly` is unaltered, but the version in the local workspace will have the new column.

To remove a column from a data frame, set its value to `NULL`. The `subset` function (see Section 6.8) can also be useful in such situations. Negative subscripts, which extract everything except those elements specified in the negative subscripts, can also be used to create a data frame with selected rows or columns removed.

Often a similar operation needs to be performed on several columns of a data frame, with the goal being to overwrite the original versions of the variables. In cases like this, the left-hand side of the assignment statement can consist of multiple columns, as long as the expression on the right-hand side is the same size as implied by the target. For example, to convert the lengths of the four numeric variables in the `iris` dataset to inches from centimeters, we can use `sapply` to operate on all four columns at once, and assign the result back to those same columns:

```
> iris[,-5] = sapply(iris[,-5],function(x)x/2.54)
```

9.2 Recoding Variables

Often times it is necessary to create a new variable based on values of an old variable. For example, in contingency table analysis we may need to group together observations with different values, and assign them all a new value. For logistic regression, it may be necessary to change a continuous variable into one that takes on values of either 0 or 1. For simple cases, logical variables can be used directly to convert a continuous variable to a binary one. For example, using the `iris` data frame, suppose we wanted to create a new variable, `bigsepal`, which would be `TRUE` when `Sepal.Length` was greater than 6, and `FALSE` otherwise. We can simply create the appropriate logical variable:

```
> bigsepal = iris$Sepal.Length > 6
```

When a logical variable is used in a numeric context, it is automatically converted to 1 if it is TRUE and 0 if it is FALSE. Thus, logical variables can be manipulated to create categorical variables with more than two levels. Suppose we wanted to create a categorical variable called sepalgroup, based on Sepal.Length, which would be equal to 1 for lengths less than or equal to 5, 2 for lengths between 5 and 7, and 3 for lengths greater than or equal to 7. We could combine logical variables as follows:

```
> sepalgroup = 1 + (iris$Sepal.Length >= 5)
+                 + (iris$Sepal.Length >= 7)
```

Note that in this case the same result could be achieved using cut (see Section 5.4):

```
> sepalgroup = cut(iris$Sepal.Length,c(0,5,7,10),
+                  include.lowest=TRUE,right=FALSE)
```

For some recoding tasks, the ifelse function may be more useful than manipulating logical variables directly. Suppose we have a variable called group that takes on values in the range of 1 to 5, and we wish to create a new variable that will be equal to 1 if the original variable is either 1 or 5, and equal to 2 otherwise. The ifelse statement accepts a logical vector as its first argument, and two other arguments: the first provides a value for the case where elements of the input logical vector are true, and the second for the case where they are false. So in this example, we could get the desired result using

```
> newgroup = ifelse(group %in% c(1,5),1,2)
```

The second and third arguments to ifelse will be recycled as necessary to be conformable with the input logical vector.

Note that the object returned by ifelse will be the same shape as the first input argument, so ifelse is effectively limited to cases where the desired result for each element is a scalar. If either of the second or third arguments to ifelse returns a vector, the return value of ifelse will be silently truncated to just its first element.

Calls to ifelse can be nested. Continuing with the previous example, if we wanted to recode values of 1 and 5 to 1, 2 and 4 to 2, and other values (in this case 3) to 3, we could use nested calls to ifelse as follows:

```
> newgroup = ifelse(group %in% c(1,5),1,
+                   ifelse(group %in% c(2,4),2,3))
```

Some words of warning about ifelse are in order. If any of the elements of the first argument to ifelse are TRUE, then all of the values in the second argument will need to be evaluated. Similarly, if any of the input elements are FALSE, then each value in the third argument must be evaluated. If either of the alternative values requires a large amount of computation, this may make ifelse surprisingly slow. Additionally, using ifelse with a variety of data may result in surprises. As a simple example, suppose we have a vector, x, and we wish to take the logarithm of the values greater than 0, and the absolute value of values less than or equal to zero. If we happen to provide a vector with all values less than zero, there is no problem:

```
> x = c(-1.2,-3.5,-2.8,-1.1,-0.7)
> newx = ifelse(x > 0,log(x),abs(x))
> newx
[1] 1.2 3.5 2.8 1.1 0.7
```

As soon as one or more values in the vector satisfy the condition x > 0, warnings will appear when R tries to evaluate the logarithm of the negative numbers, even though it will never actually return them:

```
> x = c(-1.2,-3.5,-2.8,1.1,-0.7)
> newx = ifelse(x > 0,log(x),abs(x))
Warning message:
NaNs produced in: log(x)
> newx
[1] 1.20000000 3.50000000 2.80000000 0.09531018 0.70000000
```

At the expense of some additional operations, the problem can be avoided:

```
> newx = numeric(length(x))
> newx[x > 0] = log(x[x > 0])
> newx[x <= 0] = abs(x[x <= 0])
> newx
[1] 1.20000000 3.50000000 2.80000000 0.09531018 0.70000000
```

Since expressions in R return their evaluated values, yet another solution is to use sapply with if/else expressions:

```
> newx = sapply(x,function(t)if(t > 0)log(t) else abs(t))
```

9.3 The recode Function

A very flexible approach to recoding variables is provided by the recode function of the car package, available through CRAN. Similar to facilities in other statistical languages, the recode function accepts descriptions of ranges of values along with a new, constant value to be assigned to observations within those ranges. These range/value pairs are passed to recode as a character string, with equal signs (=) separating ranges and values, and semicolons (;) separating each range/value pair.

There are four possibilities for range/value pairs:

1. single values, for example 3='control'
2. multiple values, for example c(1,5)=5

3. ranges of values, for example `5:7='middle'`. The special values `lo` and `hi` can appear in a range to represent the lowest or highest value for the variable being recoded.

4. the word `else`, representing values not covered by any other provided ranges, for example `else='not found'`.

So to recode values 1 and 5 to 1, 2 and 4 to 2, and other values to 3, we could use `recode` as follows (after loading the `car` package):

```
> newgroup = recode(group,'c(1,5)=1;c(2,4)=2;else=3')
```

9.4 Reshaping Data Frames

Often the values required for a particular operation can be found in a data frame, but they are not organized in the appropriate way. As a simple example, data for multiple groups are often stored in spreadsheets or data summaries as columns, with a separate column for each group. Most of the modeling and graphics functions in R will not be able to work with such data; they expect the values to be in a single column with an additional column that specifies the group from which the data arose. The `stack` function can reorganize datasets to have this property. As an example, suppose that data for three groups is stored in a data frame as follows:

```
> mydata = data.frame(grp1=c(12,15,19,22,25),
+                     grp2=c(18,12,42,29,44),
+                     grp3=c(8,17,22,19,31))
> mydata
  grp1 grp2 grp3
1   12   18    8
2   15   12   17
3   19   42   22
4   22   29   19
5   25   44   31
```

To perform an analysis of variance or produce histograms for each group, the data would need to rearranged using `stack`:

```
> sdata = stack(mydata)
> head(sdata)
  values  ind
1     12 grp1
2     15 grp1
3     19 grp1
4     22 grp1
5     25 grp1
6     18 grp2
```

If there were other variables in the data frame that did not need to be converted to this form, the `select=` argument to `stack` allows you to specify the variables that should be used, similar to the same argument to the `subset` function.

The `unstack` function will reorganize stacked data back to the one column per group form. To use `unstack`, a formula must be provided to explain the roles of the variables to be unstacked. To convert the `sdata` data frame back to its original form, `unstack` could be called as follows:

```
> mydata = unstack(sdata,values~ind)
> head(mydata)
  grp1 grp2 grp3
1   12   18    8
2   15   12   17
3   19   42   22
4   22   29   19
5   25   44   31
```

For more complex reorganizations, the concept of "wide" versus "long" datasets is often helpful. When there are multiple occurrences of values for a single observation, a data frame is said to be long if each occurrence is a separate row in the data frame; if all of the occurrences of values for a given observation are in the same row, then the dataset is said to be wide. The `reshape` function converts datasets between these two forms.

Perhaps the most common use of `reshape` involves repeated measures analyses, where the same variable is recorded for each observation at several different times. For some types of analysis (for example, split-plot designs), the long form is preferred; for other analyses (for example, correlation studies), the wide form is needed. For example, consider the following artificial dataset which contains observations at three different times for four subjects on variables called x and y:

```
> set.seed(17)
> obs = data.frame(subj=rep(1:4,rep(3,4)),
+                  time=rep(1:3),
+                  x=rnorm(12),y=rnorm(12))
> obs
   subj time          x           y
1     1    1 -1.01500872  1.29532187
2     1    2 -0.07963674  0.18791807
3     1    3 -0.23298702  1.59120510
                  . . .
9     3    3  0.25523700  0.68102765
10    4    1  0.36658112 -0.68203337
11    4    2  1.18078924 -0.72325674
12    4    3  0.64319207  1.67352596
```

To use `reshape` to convert the dataset to wide format, we need to provide five arguments. The first argument is the data frame to be reshaped. The next three arguments provide the names of the columns that will be involved in the reshaping. The `idvar=` argument provides the names of the variables that define the experimental unit which was repeatedly measured. In this case, it's the `subj` variable. The `v.names=` argument tells `reshape` which variables in the long format will be used to create the multiple variables in the wide format. In this example, we want both `x` and `y` be to be expanded to multiple variables, so we'd specify a vector with both those names. The `timevar=` variable tells which variable identifies the sequence number that will be used to create the multiple versions of the `v.names` variables; in this case it will be `time`. Finally, the `direction=` argument accepts values of `"wide"` or `"long"`, depending on which transformation is to be performed. Putting this all together, we can perform the conversion to wide format with the following call to `reshape`:

```
> wideobs = reshape(obs,idvar='subj',v.names=c('x','y'),
+                    timevar='time',direction='wide')
> wideobs
     subj         x.1          y.1          x.2          y.2
1       1 -1.0150087   1.29532187  -0.07963674   0.1879181
4       2 -0.8172679  -0.05517906   0.77209084   0.8384711
7       3  0.9728744   0.62595440   1.71653398   0.6335847
10      4  0.3665811  -0.68203337   1.18078924  -0.7232567
              x.3          y.3
1     -0.2329870   1.5912051
4     -0.1656119   0.1593701
7      0.2552370   0.6810276
10     0.6431921   1.6735260
```

Notice that the names of the variables are passed to `reshape`, not the actual values of the variables.

The names `x.1`, `y.1`, etc. were formed by joining together the variable names of the variables specified in the `v.names=` argument with the values of the `timevar=` variable. Any variables not specified in the `v.names=` argument are assumed to be constant for all observations with the same values as the `idvar=` variables, and a single copy of such variables will be included in the output data frame. Only the variables whose names appear in the `v.names=` argument will be converted into multiple variables, so if any variables that are in the data frame but not in the `v.names=` argument are not constant, `reshape` will print a warning message, and use the first value of such variables when converting to wide format. To prevent variables from being transferred to the output data frame, the `drop=` argument can be used to pass a vector of variable names to be ignored in the conversion.

The information about the reshaping procedure is stored as attributes in converted data frames, so once a data frame has been converted with `reshape`, it can be changed to its previous format by passing just the data frame with no additional arguments to reshape. Thus, we could convert the wideobs data frame to its original long format as follows:

```
> obs = reshape(wideobs)
> head(obs)
     subj time         x           y
1.1     1    1 -1.01500872  1.29532187
2.1     2    1 -0.81726793 -0.05517906
3.1     3    1  0.97287443  0.62595440
4.1     4    1  0.36658112 -0.68203337
1.2     1    2 -0.07963674  0.18791807
2.2     2    2  0.77209084  0.83847112
```

As an example of converting from wide to long format, consider the USPersonalExpenditure dataset. Since it is stored as a matrix, we'll first convert it to a data frame, transferring the row names into a variable called type:

```
> usp = data.frame(type=rownames(USPersonalExpenditure),
+                   USPersonalExpenditure,row.names=NULL)
> usp
                type  X1940  X1945 X1950 X1955 X1960
1      Food and Tobacco 22.200 44.500 59.60  73.2 86.80
2 Household Operation 10.500 15.500 29.00  36.5 46.20
3   Medical and Health  3.530  5.760  9.71  14.0 21.10
4         Personal Care  1.040  1.980  2.45   3.4  5.40
5    Private Education  0.341  0.974  1.80   2.6  3.64
```

Since reshape can handle multiple sets of variables, the varying= argument should be passed a list containing vectors with the names of the different sets of variables that should be mapped to a single variable in the long dataset. In the current example, there is only one set of variables to be mapped, so we pass a list with a vector of the appropriate variable names. Along with the direction='long' argument, this list will usually be enough to convert the dataset:

```
> rr = reshape(usp,varying=list(names(usp)[-1]),direction='long')
> head(rr)
                type time   X1940 id
1.1      Food and Tobacco    1  22.200  1
2.1  Household Operation    1  10.500  2
3.1    Medical and Health    1   3.530  3
4.1         Personal Care    1   1.040  4
5.1    Private Education    1   0.341  5
1.2      Food and Tobacco    2  44.500  1
```

By providing additional information to reshape, the resulting data frame can be modified to provide more useful information. For example, the automatically generated variable id is simply a numeric index corresponding to the `type` variable; using `idvar='type'` will suppress its creation. The automatically generated variable `time` defaults to a set of consecutive integers; providing more meaningful values through the `times=` argument will label the values properly. Finally, the name of the column representing the values (which defaults to the first name in the `varying=` argument) can be set to a more meaningful name with the `v.names=` argument.

```
> rr=reshape(usp,varying=list(names(usp)[-1]),idvar='type',
+           times=seq(1940,1960,by=5),v.names='expend',
+           direction='long')
> head(rr)
                          type time expend
Food and Tobacco.1940        Food and Tobacco 1940 22.200
Household Operation.1940 Household Operation 1940 10.500
Medical and Health.1940     Medical and Health 1940  3.530
Personal Care.1940             Personal Care 1940  1.040
Private Education.1940     Private Education 1940  0.341
Food and Tobacco.1945        Food and Tobacco 1945 44.500
```

In cases like this, where the desired value for time is embedded in the variable names being converted, the `split=` argument can be used to automatically determine the values for the times and names for the variables containing the values. When you use the `split=` argument, the `varying=` argument should be a vector, not a list, because `reshape` will figure out the sets of variables based on the prefixes found by splitting the variable names. The `split=` argument is passed as a list with two elements: `regexp` and `include`. The `regexp` argument provides a regular expression used to split up the names provided through the `varying=` argument. The first split piece will be used as a name for the variable containing the values, and the second split piece will be used to form values for the `time` variable that `reshape` generates. To keep the regular expression as part of the names and values that are created, the `include` argument should be set to `TRUE`. So an alternative way of reshaping the `usp` data frame, without having to explicitly provide the values of the times, would be:

```
> rr1 = reshape(usp,varying=names(usp)[-1],idvar='type',
+          split=list(regexp='X1',include=TRUE),direction='long')
> head(rr1)
                          type time      X
Food and Tobacco.1940        Food and Tobacco 1940 22.200
Household Operation.1940 Household Operation 1940 10.500
Medical and Health.1940     Medical and Health 1940  3.530
Personal Care.1940             Personal Care 1940  1.040
Private Education.1940     Private Education 1940  0.341
Food and Tobacco.1945        Food and Tobacco 1945 44.500
```

To replace the generated row names with ones of your own choosing, use the `new.row.names=` argument.

9.5 The reshape Package

The reshape package, introduced in Section 8.6, uses the concept of "melting" a dataset (through the melt function) into a data frame which contains separate columns for each id variable, a variable column containing the name of each measured variable, and a final column named value with the variable's value. It may be noticed that this melting operation is essentially a "wide-to-long" reshaping of the data. Using the usp data frame from a previous example, we can easily convert the melted form to the long form as follows:

```
> library(reshape)
> usp = data.frame(type=rownames(USPersonalExpenditure),
+                  USPersonalExpenditure,row.names=NULL)
> musp = melt(usp)
> head(musp)
                 type variable  value
1      Food and Tobacco    X1940 22.200
2 Household Operation    X1940 10.500
3  Medical and Health    X1940  3.530
4         Personal Care    X1940  1.040
5     Private Education    X1940  0.341
6      Food and Tobacco    X1945 44.500
```

To complete the conversion, we need only remove the "X" from the variable column, rename it to time, and rename the value column to expend:

```
> musp$variable = as.numeric(sub('X','',musp$variable))
> names(musp)[2:3] = c('time','expend')
> head(musp)
                 type time expend
1      Food and Tobacco 1940 22.200
2 Household Operation 1940 10.500
3  Medical and Health 1940  3.530
4         Personal Care 1940  1.040
5     Private Education 1940  0.341
6      Food and Tobacco 1945 44.500
```

Keep in mind that variable is a factor, and that the sub function converts it to a character before operating on it; if you use it directly, you may need to pass it to as.character before processing. Since both the id variables and measure variables appear in the columns of the "long" dataset, this transformation could also be performed using

```
cast(musp,variable + type ~ .)
```

For long-to-wide conversions, recall that variables appearing to the left of the tilde in the formula passed to cast will appear in the columns of the output, while those on the right will appear in the rows. Using the simulated data from the previous section, we put `subj` on the left-hand side of the formula and `variable` (created by the `melt` function) and `time` on the right:

```
> set.seed(17)
> obs = data.frame(subj=rep(1:4,rep(3,4)),
+                  time=rep(1:3),
+                  x=rnorm(12),y=rnorm(12))
> mobs = melt(obs)
> cast(subj ~ variable + time,data=mobs)
  subj        x_1         x_2        x_3        y_1        y_2
1    1 -1.0150087 -0.07963674 -0.2329870  1.29532187  0.1879181
2    2 -0.8172679  0.77209084 -0.1656119 -0.05517906  0.8384711
3    3  0.9728744  1.71653398  0.2552370  0.62595440  0.6335847
4    4  0.3665811  1.18078924  0.6431921 -0.68203337 -0.7232567
        y_3
1 1.5912051
2 0.1593701
3 0.6810276
4 1.6735260
```

The names of the derived columns are constructed in the order in which the right-hand-side variables are entered in the formula.

To separate each time into a separate list element, the vertical bar (|) can be used:

```
> cast(subj ~variable|time,data=mobs)
$'1'
  subj          x          y
1    1 -1.0150087  1.29532187
2    2 -0.8172679 -0.05517906
3    3  0.9728744  0.62595440
4    4  0.3665811 -0.68203337

$'2'
  subj           x          y
1    1 -0.07963674  0.1879181
2    2  0.77209084  0.8384711
3    3  1.71653398  0.6335847
4    4  1.18078924 -0.7232567

$'3'
  subj          x         y
1    1 -0.2329870 1.5912051
2    2 -0.1656119 0.1593701
3    3  0.2552370 0.6810276
4    4  0.6431921 1.6735260
```

It can be noted that this performs the same operation as the split function (Section 8.5), but the redundant variable (time in this example) is not included in the output.

Remember that the dataset that cast is operating on is the melted dataset, not the original one. So to create a wide data frame from the simulated data, but only including x, we could use

```
> cast(subj ~ variable + time,subset = variable == 'x',data=mobs)
  subj       x_1         x_2         x_3
1    1 -1.0150087 -0.07963674 -0.2329870
2    2 -0.8172679  0.77209084 -0.1656119
3    3  0.9728744  1.71653398  0.2552370
4    4  0.3665811  1.18078924  0.6431921
```

9.6 Combining Data Frames

At the most basic level, two or more data frames can be combined by rows using rbind, or by columns using cbind. For rbind, the data frames must have the same number of columns; for cbind, the data frames must have the same number of rows. Vectors or matrices passed to cbind will be converted to data frames, so the mode of columns passed to cbind will be preserved.

While cbind will demand that data frames and matrices are conformable (that is, they have the same number of rows), vectors passed to cbind will be recycled if the number of rows in the data frame or matrix is an even multiple of the length of the vector. Consider the following two data frames, one with three rows, and one with four:

```
> x = data.frame(a=c('A','B','C'),x=c(12,15,19))
> y = data.frame(a=c('D','E','F','G'),x=c(19,21,14,12))
```

We can bind a vector with two values to the second data frame, since four is an even multiple of two; R will recycle the vectors values to insure conformability:

```
> cbind(y,z=c(1,2))
  a  x z
1 D 19 1
2 E 21 2
3 F 14 1
4 G 12 2
```

When using cbind, duplicate column names will not be detected:

```
> cbind(x,y[1:3,])
  a  x a  x
1 A 12 D 19
2 B 15 E 21
3 C 19 F 14
```

It may be a good idea to use unique names when combining data frames in this way. An easy way to test is to pass the names of the two data frames to the `intersect` function:

```
> intersect(names(x),names(y))
[1] "a" "x"
```

When using `rbind`, the names and classes of values to be joined must match, or a variety of errors may occur. This is especially important when values in any of the columns involved are factors. Using the `data.frame` function when adding rows to a data frame can usually resolve the problem:

```
> z = rbind(x,c(a='X',x=12))
Warning message:
invalid factor level, NAs generated in:
"[<-.factor"('*tmp*', ri, value = "X")
> z = rbind(x,data.frame(a='X',x=12))
> levels(z$a)
[1] "A" "B" "C" "X"
```

Although the `rbind` function will demand that the names of the objects being combined agree, `cbind` does not do any such checking. To combine data frames based on the values of common variables, the `merge` function should be used. This function is designed to provide the same sort of functionality and behavior as the table joins provided by relational databases. Although `merge` is limited to operating on two data frames at a time, it can be called repeatedly to deal with more than two data frames.

The default behavior of `merge` is to join together rows of the data frames based on the values of all of the variables (columns) that the data frames have in common. (In database terminology, this is known as a natural join.) When called without any other arguments, `merge` returns only those rows which had observations in both data frames. As a simple example, consider the merge resulting from these two data frames, each of which has rows with values of the merging variable that are not found in the other data frame:

```
> x = data.frame(a=c(1,2,4,5,6),x=c(9,12,14,21,8))
> y = data.frame(a=c(1,3,4,6),y=c(8,14,19,2))
> merge(x,y)
  a  x  y
1 1  9  8
2 4 14 19
3 6  8  2
```

Although there were six unique values for a between the two data frames, only those rows with values of a in both data frames are represented in the output. To modify this, the `all=`, `all.x=`, and `all.y=` arguments can be used. Specifying `all=TRUE` will include all rows (full outer join, in database terminology), `all.x=TRUE` will include all rows from the first data frame (left outer join), and `all.y=TRUE` does the same for the second data frame (right outer join). Each case can be illustrated with the current example:

```
> merge(x,y,all=TRUE)
  a  x  y
1 1  9  8
2 2 12 NA
3 3 NA 14
4 4 14 19
5 5 21 NA
6 6  8  2
> merge(x,y,all.x=TRUE)
  a  x  y
1 1  9  8
2 2 12 NA
3 4 14 19
4 5 21 NA
5 6  8  2
> merge(x,y,all.y=TRUE)
  a  x  y
1 1  9  8
2 3 NA 14
3 4 14 19
4 6  8  2
```

Note that missing values (NA) are inserted in the places where data was missing from one of the data frames.

To take more control over which variables are used to merge rows of the data frame, the by= argument can be used. You provide the by= argument with a vector of the name or names of the variables that should be used for the merge. If the merging variables have different names in the data frames to be merged, the by.x= and by.y= arguments can be used.

When there are multiple rows with common values of the merging variable in either of the data frames being merged, each row will contribute one observation to the output data frame. If one of the datasets has exactly one observation for each value of the merging variable(s), the resultant merge is sometimes known as a table lookup. As a simple example, consider two datasets, one with city names and state abbreviations, and a second with state abbreviations and full state names. The goal is to create a dataset with the names of the cities along with the full state names. The following dataset represents 10 of the most expensive cities in the United States, based on housing and food costs:

```
> cities = data.frame(city=c('New York','Boston','Juneau',
+                           'Anchorage','San Diego',
+                           'Philadelphia','Los Angeles',
+                           'Fairbanks','Ann Arbor','Seattle'),
+                   state.abb= c('NY','MA','AK','AK','CA',
+                                 'PA','CA','AK','MI','WA'))
```

```
> cities
            city state.abb
1       New York        NY
2         Boston        MA
3         Juneau        AK
4      Anchorage        AK
5      San Diego        CA
6   Philadelphia        PA
7    Los Angeles        CA
8      Fairbanks        AK
9      Ann Arbor        MI
10       Seattle        WA
```

A corresponding data frame with state abbreviations and full names can be formed as follows:

```
> states = data.frame(state.abb= c('NY','MA','AK','CA',
+                               'PA','MI','WA'),
+            state=c('New York','Massachusetts','Alaska',
+                    'California','Pennsylvania',
+                    'Michigan','Washington'))
```

Note that there is exactly one observation for each state/abbreviation combination in the states dataset. With this restriction in place, merging the two datasets is simple (since they have a single variable, state.abb, in common:

```
> merge(cities,states)
    state.abb          city          state
1         AK        Juneau         Alaska
2         AK     Anchorage         Alaska
3         AK     Fairbanks         Alaska
4         CA     San Diego     California
5         CA   Los Angeles     California
6         MA        Boston  Massachusetts
7         MI     Ann Arbor       Michigan
8         NY      New York       New York
9         PA  Philadelphia   Pennsylvania
10        WA       Seattle     Washington
```

The multiple observations per state in the cities data frame cause no problem, because there was always exactly one matching observation in the states data frame.

Now suppose we (foolishly) create a data frame with the zip codes for various cities using only the state abbreviation as an identifier. The problem is that there will be more than one zip code for some of the states, making it impossible for merge to know exactly which observations should be joined together. In cases like this, merge silently creates multiple observations so

that there will be an observation for each multiple occurrence of the merging
variables in the merged data frame.

```
> zips = data.frame(state.abb=c('NY','MA','AK','AK','CA',
+                               'PA','CA','AK','MI','WA'),
+          zip=c('10044','02129','99801','99516','92113',
+                '19127','90012','99709','48104','98104'))
> merge(cities,zips)
   state.abb        city   zip
1         AK      Juneau 99801
2         AK      Juneau 99516
3         AK      Juneau 99709
4         AK   Anchorage 99801
5         AK   Anchorage 99516
6         AK   Anchorage 99709
7         AK   Fairbanks 99801
8         AK   Fairbanks 99516
9         AK   Fairbanks 99709
10        CA   San Diego 92113
11        CA   San Diego 90012
12        CA Los Angeles 92113
13        CA Los Angeles 90012
14        MA      Boston 02129
15        MI   Ann Arbor 48104
16        NY    New York 10044
17        PA Philadelphia 19127
18        WA     Seattle 98104
```

Now there are 18 observations in the output dataset instead of the expected
10. For any state for which there were multiple observations in the `zips`
data frame, `merge` has created that many observations for each observation
in the `cities` dataset with that value of `state.abb`. The moral is that you
should proceed with caution when you have multiple occurrences of values of
a merging variable in both of the datasets being merged.

9.7 Under the Hood of merge

While the `merge` function will perform most common tasks regarding combining two data frames, it is occasionally useful to just find the indexes of
common values in two vectors, rather than actually combining them. Internally, `merge` uses the `match` function to find these indexes. This function
requires two arguments: the first is a vector of values to be matched, and the
second is the vector of values that should be searched for possible matches.
For those elements in the first vector that had matching values in the second,

match returns the index of the first such value in the second vector; for elements that didn't match, the default behavior of match is to return a missing value (NA). Thus, the return value from match will always be a vector of the same length as the first argument. For example, in Section 9.6, we merged the cities and states data frames based on common values of the state.abb variable. To retrieve just the indexes of the matching values, we could call match as follows:

```
> match(cities$state.abb,states$state.abb)
 [1] 1 2 3 3 4 5 4 3 6 7
```

The nomatch= argument can be used to provide a different value to be returned when a match was not found. Since subscripts of 0 are ignored, one very useful choice for this value is nomatch=0. When this value is used, the result from match can be used as an index to the second vector to find the values that actually matched. Continuing with the x and y example from Section 9.6, suppose we wanted to know which values in x$a were also present in x$b. By calling match with nomatch=0, the resulting vector can be used as an index into y$b to extract the actual values:

```
> indices = match(x$a,y$a,nomatch=0)
> y$a[indices]
[1] 1 4 6
```

It may be noted that this is equivalent to the intersect function, which currently uses match to do its work.

Finally, for the simpler case where interest is only in whether or not elements in one vector can be found in another vector, the %in% operator can be used. To produce a logical vector, the same length as x$a, which indicates which values could be found in y$a, we can use %in% as follows:

```
> x$a %in% y$a
[1]  TRUE FALSE  TRUE FALSE  TRUE
```

Like intersect, %in% is currently defined using match.

Index

springer.com

Interactive and Dynamic Graphics For Data Analysis

Dianne Cook and Deborah F. Swayne

This richly illustrated book describes the use of interactive and dynamic graphics as part of multidimensional data analysis. Chapters include clustering, supervised classification, and working with missing values. A variety of plots and interaction methods are used in each analysis, often starting with brushing linked low-dimensional views and working up to manual manipulation of tours of several variables.

2007, Approx. 205 pp Softcover ISBN 978-0-387-71761-6

Graphics of Large Datasets Visualizing a Million

Antony Unwin, Martin Theus, and Heike Hoffman

This book shows how to look at ways of visualizing large datasets, whether large in numbers of cases, or large in numbers of variables, or large in both. All ideas are illustrated with displays from analyses of real datasets and the importance of interpreting displays effectively is emphasized. Graphics should be drawn to convey information and the book includes many insightful examples. The book is accessible to readers with some experience of drawing statistical graphics.

2006, XXII 271 pp. Hardcover ISBN 978-0-387-32906-2

Bayesian Computation with R

Antony Unwin, Martin Theus, and Heike Hoffman

This book introduces Bayesian modeling by the use of computation using the R language. Bayesian computational methods such as Laplace's method, rejection sampling, and the SIR algorithm are illustrated in the context of a random effects model. The construction and implementation of Markov Chain Monte Carlo (MCMC) methods is introduced. These simulation-based algorithms are implemented for a variety of Bayesian applications such as normal and binary response regression, hierarchical modeling, order-restricted inference, and robust modeling.

2007, X, 267 pp. Softcover ISBN 978-0—387-71384-7